Grade 3

Unit 4

Weather Impacts

To obtain permission(s) or for inquiries, submit a request to:

Discovery Education, Inc.
4350 Congress Street, Suite 700
Charlotte, NC 28209
800-323-9084
Education_Info@DiscoveryEd.com

ISBN 13: 978-1-68220-797-0

Printed in the United States of America.

1 2 3 4 5 LBC 28 27 26 25 24 A

Acknowledgments

Acknowledgment is given to photographers, artists, and agents for permission to feature their copyrighted material.

Cover and inside cover art: Kevin Key / Shutterstock.com

Table of Contents

Unit 4: Weather Impacts

Concept 4.3 Weather Hazards

Unit Wrap-Up

Grade 3 Resources

Dear Parent/Guardian,

This year, your student will be using Science Techbook™, a comprehensive science program developed by the educators and designers at Discovery Education and written to the Next Generation Science Standards (NGSS). The NGSS expect students to act and think like scientists and engineers, to ask questions about the world around them, and to solve real-world problems through the application of critical thinking across the domains of science (Life Science, Earth and Space Science, Physical Science).

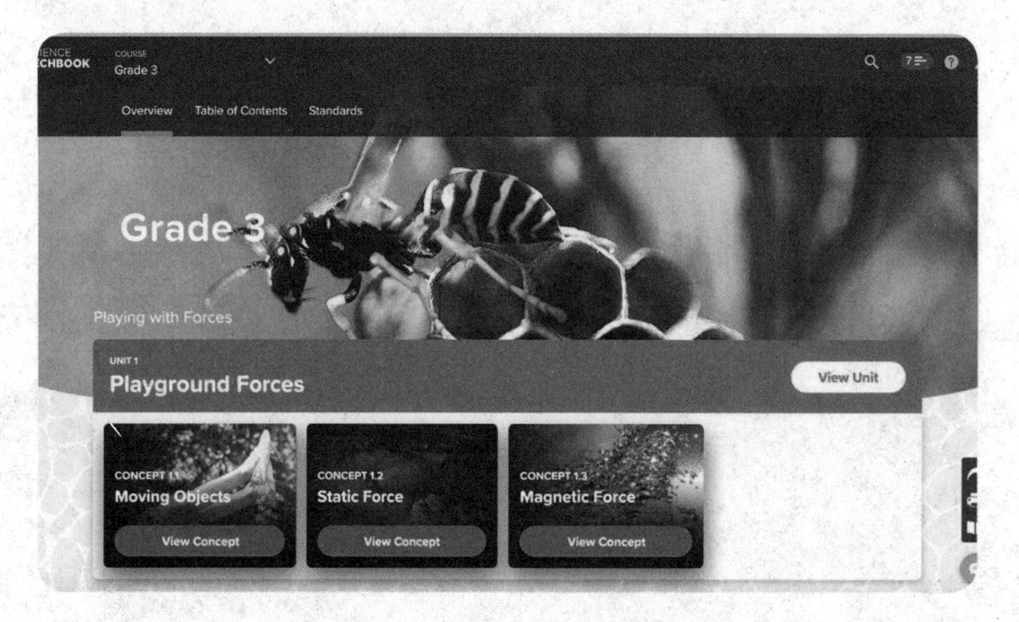

Science Techbook is an innovative program that helps your student master key scientific concepts. Students engage with interactive science materials to analyze and interpret data, think critically, solve problems, and make connections across science disciplines. Science Techbook includes dynamic content, videos, digital tools, Hands-On Activities and labs, and game-like activities that inspire and motivate scientific learning and curiosity.

You and your child can access the resource by signing in to www.discoveryeducation.com. You can view your child's progress in the course by selecting the Assignment button.

Science Techbook is divided into units, and each unit is divided into concepts. Each concept has three sections: Wonder, Learn, and Share.

Units and Concepts Students begin to consider the connections across fields of science to understand, analyze, and describe real-world phenomena.

Wonder Students activate their prior knowledge of a concept's essential ideas and begin making connections to a real-world phenomenon and the **Can You Explain?** question.

Learn Students dive deeper into how real-world science phenomenon works through critical reading of the Core Interactive Text. Students also build their learning through Hands-On Activities and interactives focused on the learning goals.

Share Students share their learning with their teacher and classmates using evidence they have gathered and analyzed during Learn. Students connect their learning with STEM careers and problem-solving skills.

Within this Student Edition, you'll find QR codes and quick codes that take you and your student to a corresponding section of Science Techbook online. To use the QR codes, you'll need to download a free QR reader. Readers are available for phones, tablets, laptops, desktops, and other devices. Most use the device's camera, but there are some that scan documents that are on your screen.

For resources in Science Techbook, you'll need to sign in with your student's username and password the first time you access a QR code. After that, you won't need to sign in again, unless you log out or remain inactive for too long.

We encourage you to support your student in using the print and online interactive materials in Science Techbook, on any device. Together, may you and your student enjoy a fantastic year of science!

Sincerely,

The Discovery Education Science Team

Discovery
EDUCATION

Unit 4

Weather Impacts

Flooded Farm

Quick Code:
us3756s

Do you like the rain? What if it lasted for several days? Rain is one way an area can get flooded, especially if there is a river nearby. This farm was flooded. Maybe the residents were able to predict the rains that brought the floods and evacuate to higher ground. Can you think of something they could have built to keep their house safe?

Flooded Farm

Think About It

Look at the photograph. **Think** about the following questions.

- How can the weather be predicted?
- How does climate change around the world?
- What solutions exist for severe weather events?

Storm

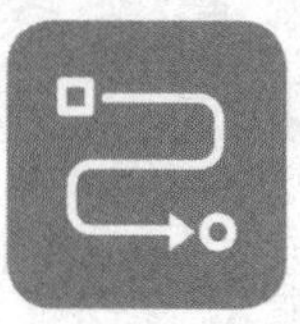

Solve Problems Like a Scientist

Quick Code: us3757s

Unit Project: Mudslides and Floods

In this project, you will use what you know about weather to design solutions to flooding and mudslides.

Mudslides and Floods

- **SEP** Engaging in Argument from Evidence
- **SEP** Asking Questions and Defining Problems
- **SEP** Constructing Explanations and Designing Solutions
- **CCC** Cause and Effect

Ask Questions About the Problem

You are going to design a barrier to stop water or mud from flowing to a certain area. **Write** some questions you can ask to learn more about the problem. As you learn about weather hazards in this unit, **write** down answers to your questions.

CONCEPT

4.1

Regional Climates

Student Objectives

By the end of this lesson:

- ☐ I can explain the pattern of Earth's climates and the predictable changes that occur every year in the climates.
- ☐ I can make a model that describes what things cause Earth to have different climates and how those things can change Earth's climates over long periods of time.
- ☐ I can use evidence to show that distance from a coast will affect the climate of an area.
- ☐ I can observe large-scale climate patterns to predict smaller-scale weather conditions.

Key Vocabulary

- ☐ atmosphere
- ☐ climate
- ☐ coast
- ☐ equator
- ☐ forecast
- ☐ heat
- ☐ humidity
- ☐ precipitation
- ☐ predict
- ☐ rain
- ☐ region
- ☐ water
- ☐ weather

Quick Code: us3755s

Activity 1

Can You Explain?

How does knowing the climate of different regions help predict the type of weather that a region generally experiences?

Quick Code:
us3759s

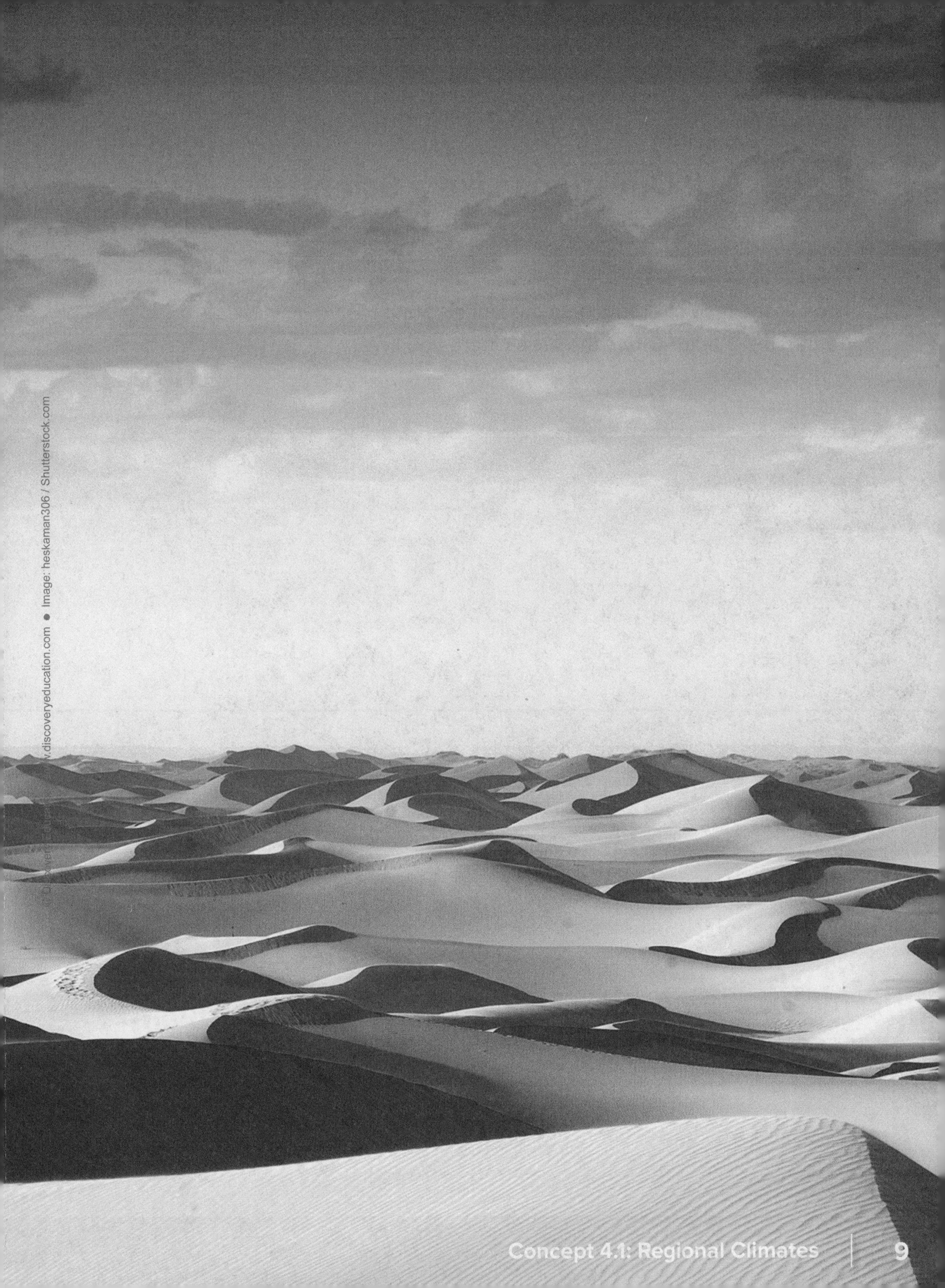

Activity 2

Ask Questions Like a Scientist

Quick Code:
us3760s

Droughts

Look at the map of droughts in the United States. Is your home on the map? **Circle** the area where you live.

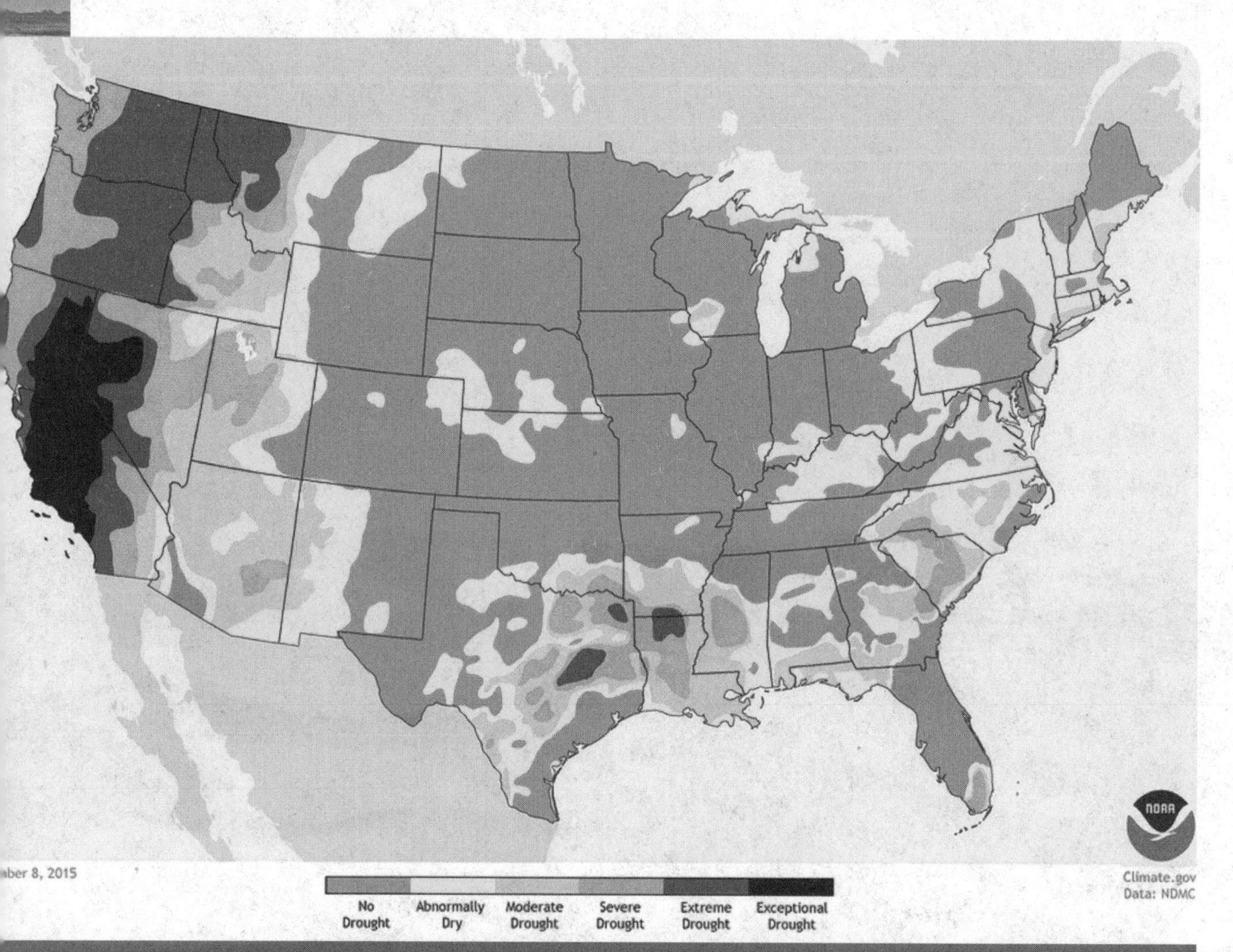

Let's Investigate Droughts

What do you wonder about the patterns on this map? **Write** three questions you have and **share** them with the class.

I wonder...

I wonder...

I wonder...

Activity 3

Analyze Like a Scientist

Quick Code: us3761s

Factors That Affect Climate

Read the text. As you read, **record** important details in the Bubble Map. You can find this map in the Share section. Then, **answer** the questions that follow.

Factors That Affect Climate

What is the **weather** like today? Is it usually like this at this time of year? In this concept, we will be looking at long-term patterns of weather and what affects them.

Do you know anyone who lives in another country? Have you studied other countries in school? Why do different places around the world have different weather patterns? What are some ways that your **climate** affects how you live? If you live in the mountains, you might need a coat much of the year. But if you live in the southern United States, you can probably wear a T-shirt without a coat—even in winter! What causes climates to be different? How do these differences impact the people who live there?

In this concept, you will learn about different climates. You will learn where they are found, and you will find out more about the typical weather in these climate zones.

Desert

Discuss with Your Class

What is climate?

____

____

____

____

What affects climate?

What is the local climate like where you live? How is it the same as or different from the climate shown in the picture?

What is climate change?

Activity 4

Observe Like a Scientist

Quick Code:
us3762s

Global Climate Zones

Watch the video. **Look** for details about the different types of climates that exist in the world.

Global Climate Zones

Talk Together

Now, talk together about the different climates shown in the video. Where are they located? What would happen if the climate in an area changed?

Activity 5

Evaluate Like a Scientist

Quick Code:
us3763s

What Do You Already Know About Regional Climates?

Influence on Climate

Look at the scenes in each of the images below. Many things can affect climate. **Circle** each of the images that shows something that affects climate. It is okay if you do not know all of the answers yet.

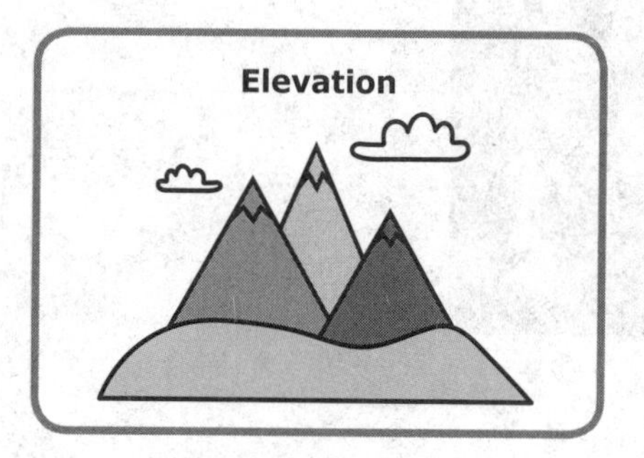

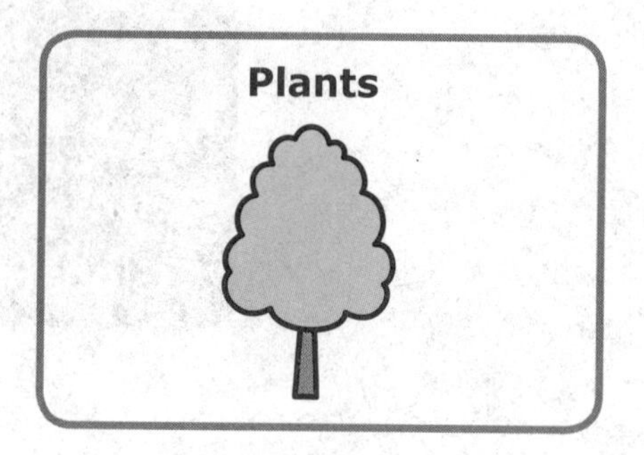

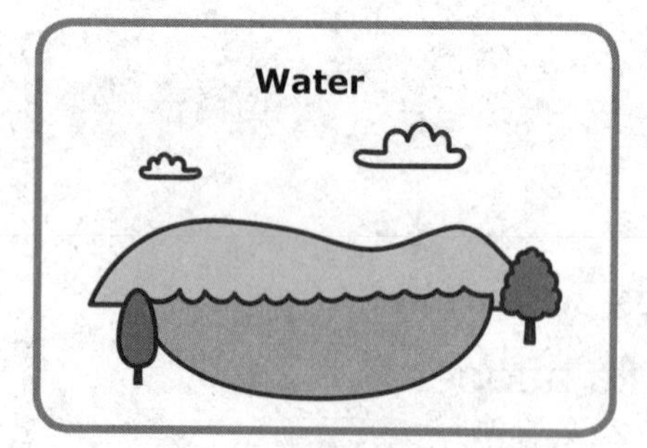

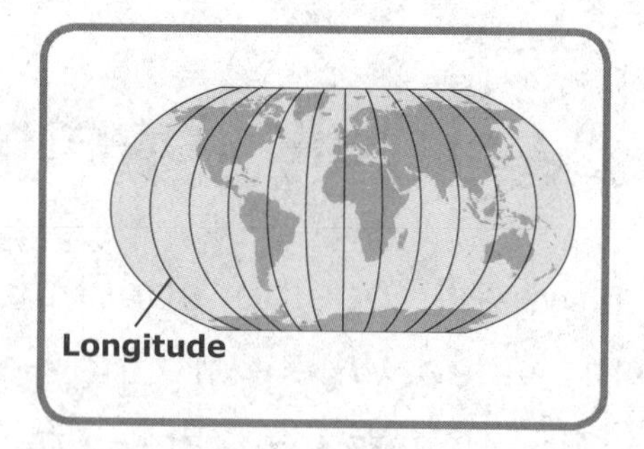

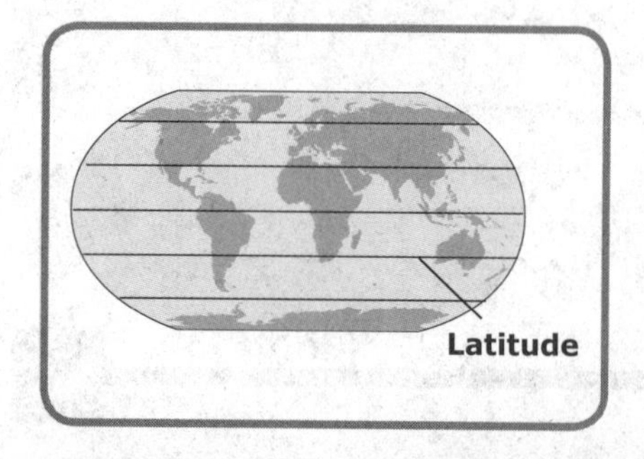

What Affects Weather and Climate?

What are three factors that can affect the climate of a region? **Write** or **draw** your answers.

What is one factor that does NOT affect the climate of a region? **Write** or **draw** your answer.

Weather and Climate

Decide whether each of the words or phrases below describes or affects weather, climate, or both weather and climate. **Use** the pictures to help you complete the activity.

Write each word or phrase in the correct circle of the diagram.

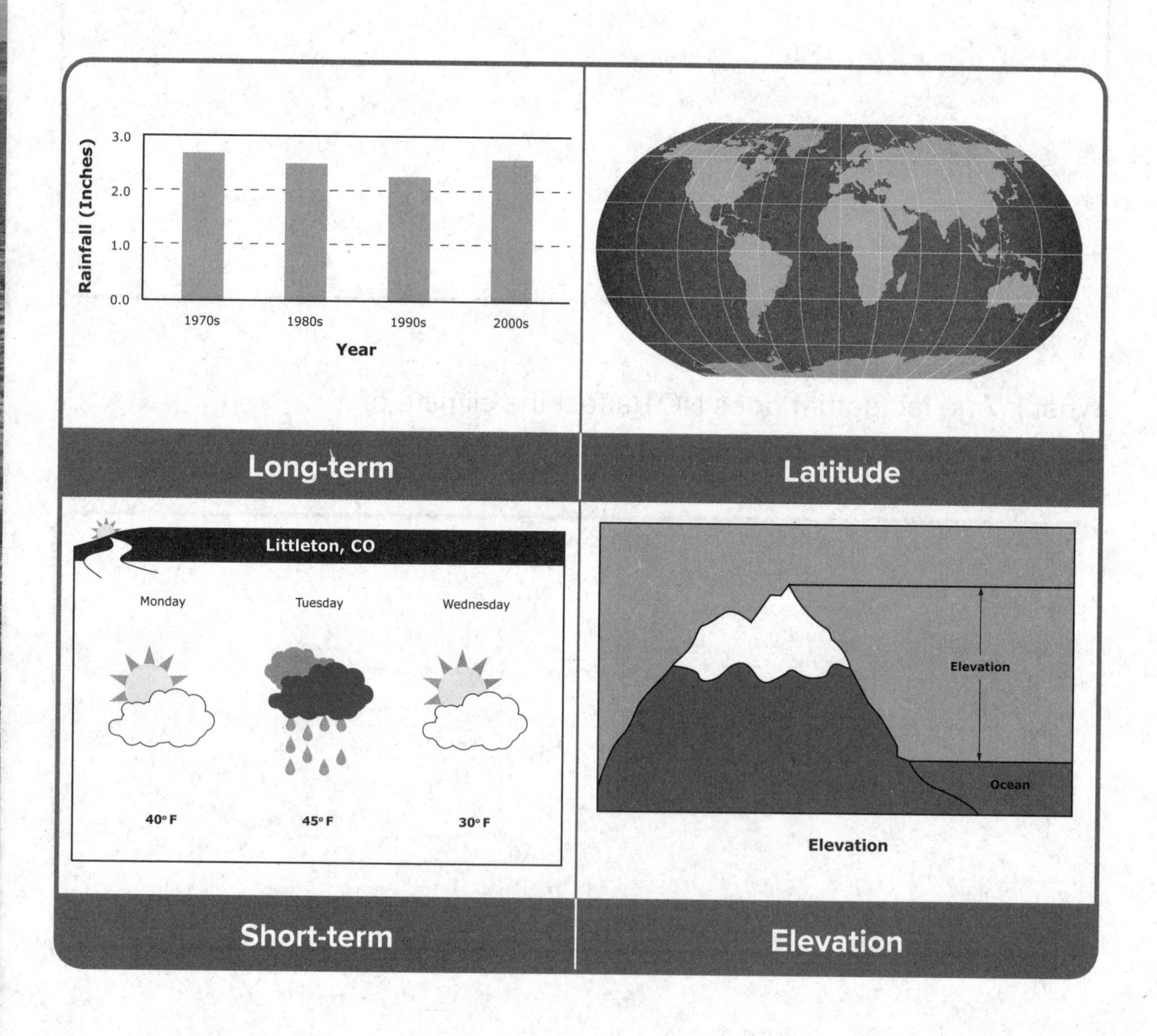

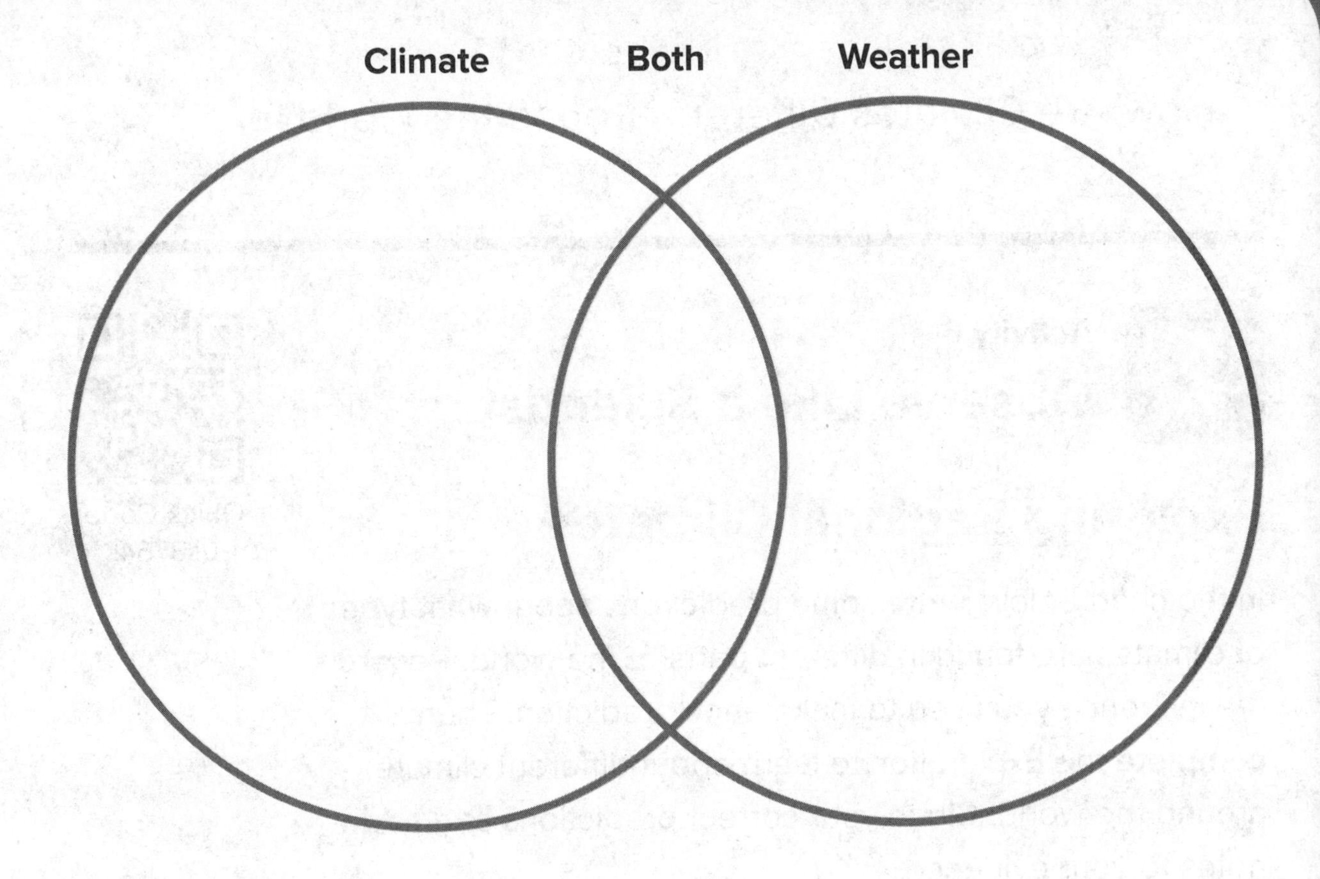
Climate
Both
Weather

How Are Climates Different around the World?

Activity 6

Observe Like a Scientist

Quick Code: us3764s

Exploring Regional Climates

In the chart below, **write** some predictions about what types of climates are found in different parts of the world. **Record** the evidence you used to make each prediction. Then, **complete** the Exploration to learn about different climates around the world. **Circle** your correct predictions and **add** notes to your evidence.

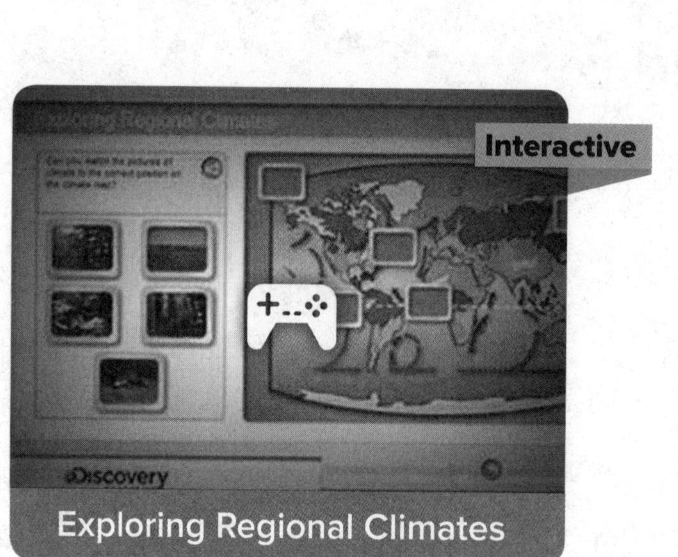

Exploring Regional Climates

SEP **Engaging in Argument from Evidence**

My Predictions	My Evidence

Activity 7

Analyze Like a Scientist

Quick Code:
us3765s

How Are Climates Different around the World?

Read the text and **watch** the video. Then, **complete** the activity that follows.

How Are Climates Different around the World?

An area's climate is determined by its average weather conditions. These conditions include temperature, **precipitation**, and other factors such as elevation, distance from a large body of **water**, ocean currents, and winds.

What Is Climate?

Different parts of the world have different climates. Some climates are hot and dry. Others are cold and wet. All have some type of seasons, but those seasons may be different from what you experience

SEP **Obtaining, Evaluating, and Communicating Information**
CCC **Patterns**

where you live. Climates are different at the **equator** than at the poles. This is because light from the sun strikes Earth's surface differently at different latitudes. The climate where you live may be different from climates in other parts of the world.

Work with a group to **write** a song about the climate where you live. Use details from what you have read and watched. To plan your song, **write** or **draw** some details about your climate in the chart.

Temperature	
Precipitation	
Seasons	

Activity 8

Evaluate Like a Scientist

Quick Code: us3766s

Understanding Climate Zones

Climate Zones

Read the list of climate characteristics below. Then, **write** the characteristics on the map, in the climate zone where they typically occur. Characteristics can be written in more than one region. Some characteristics may not appear in any regions.

- Very cool summers
- Warm summers
- Warm winters
- Very cold winters
- Wet season
- Dry season
- One season

CCC **Cause and Effect**

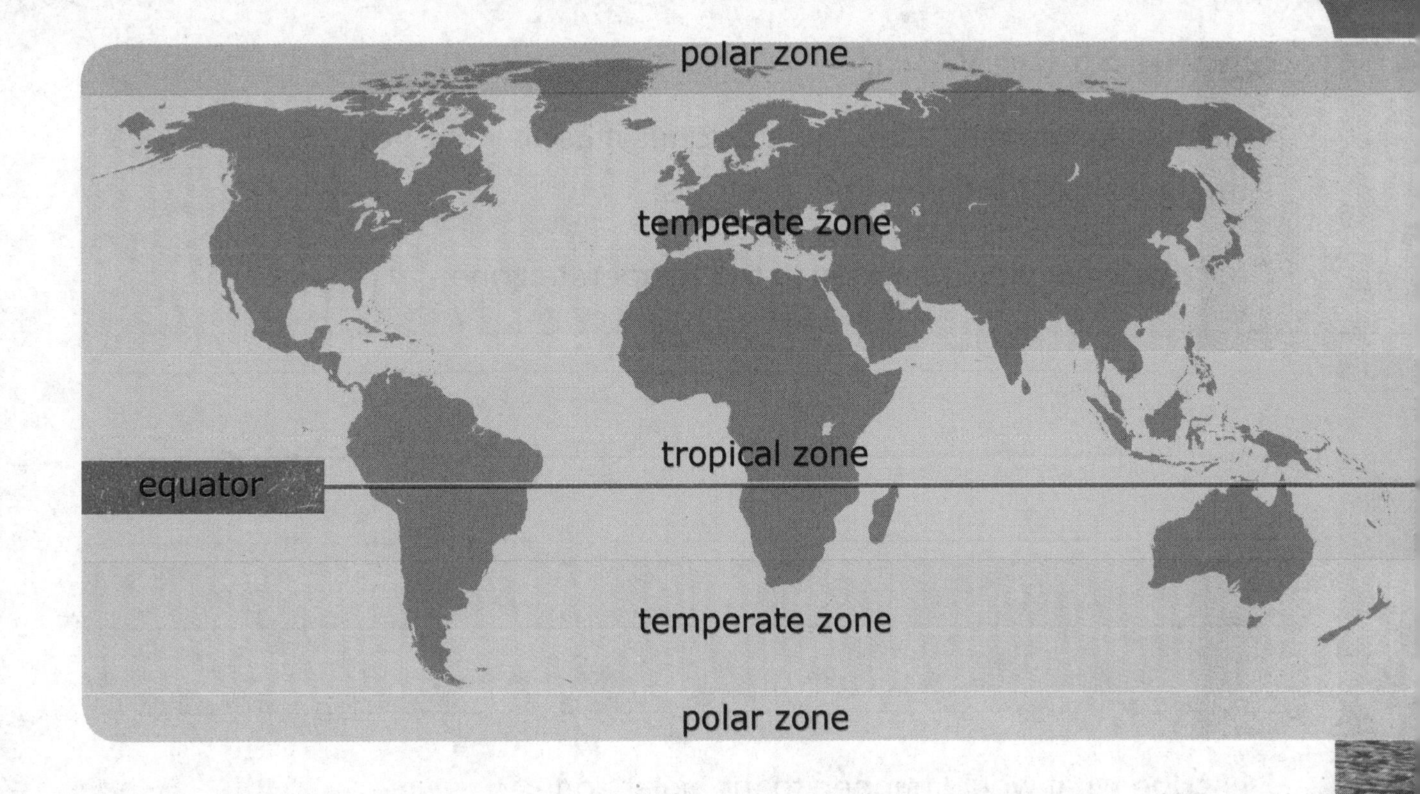
polar zone
temperate zone
equator
tropical zone
temperate zone
polar zone

Changing the World

What would happen to the climate zones if Earth were not tilted on its axis? **Answer** the questions.

Describe what would happen to the temperate zone.

Describe what would happen to the polar zone.

What Is the Climate Like in Different Areas around the World?

Activity 9

Observe Like a Scientist

Quick Code:
us3767s

Desert Biome

All deserts get very little rain each year. You might picture a desert as being hot and dry, maybe even having large sand dunes. However, deserts occur in cold regions, too.

Watch the videos to learn more about deserts. Then, **answer** the questions.

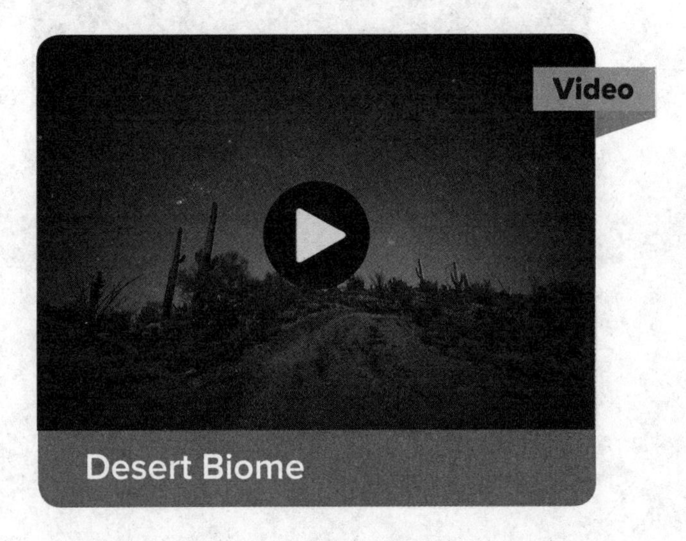

Deserts are found on almost every continent. How do deserts differ based on where they are located?

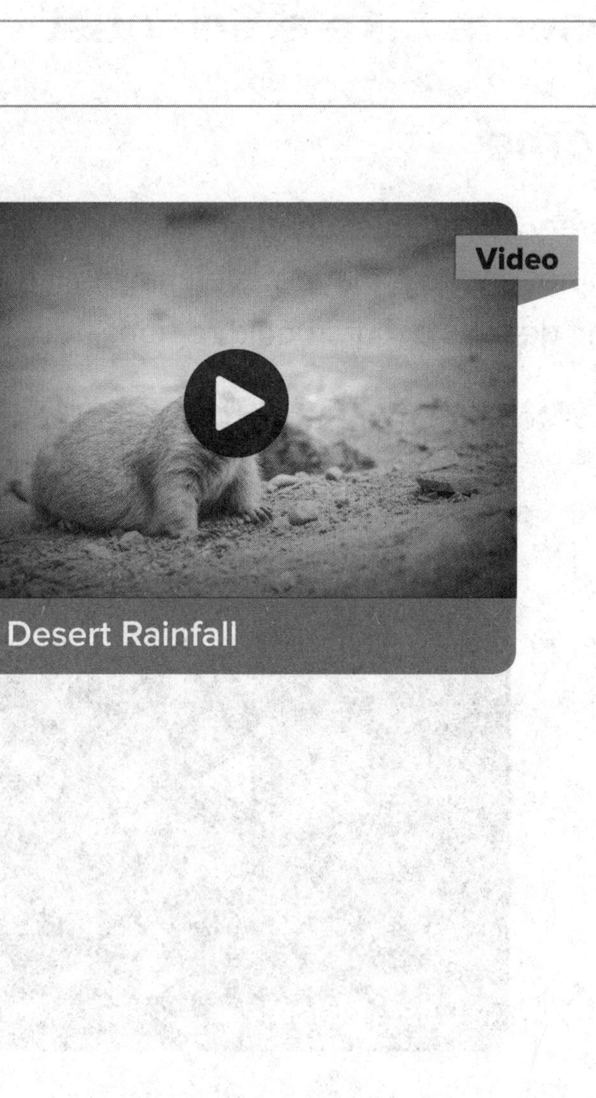

Desert Rainfall

What happens to water in the desert?

Where are the largest deserts in the world located?

Activity 10

Analyze Like a Scientist

Quick Code:
us3768s

Coastal Areas

Read the text and **watch** the video. Then, **answer** the question.

Coastal Areas

The **coast** is an area where the ocean meets the land. The world has coasts near the warm equator and the cold Arctic. Coastal regions can be hot, cold, and everything in between. They also experience different amounts of rainfall based on their location.

How is an area's climate affected if it is located on a coast? The ocean has currents that move **water** around the globe, bringing different temperatures. A warm current can bring warm water from the equator to the poles. A warm current can also heat the air around it, which will bring **humidity** inland. Cold currents can cause cool, dry air that can contribute to a desert climate. Currents often follow the same patterns, and they contribute to the climate of a region.

There are coasts all over the world, and they have different climates. What types of weather do you see at different coasts?

Activity 11

Analyze Like a Scientist

Quick Code:
us3769s

Climate Research

Explore more about the climates in three different regions of the world. **Read** the text, **watch** the videos, and **answer** the questions. Then, **choose** one of the regions and do more research on it.

Climate Research

As the name suggests, polar climates occur at or near the North and South Poles. These regions are cold all year long. Because Earth is tilted on its axis, the poles have long, light summers and long, dark winters.

Polar Habitats

At extreme northern and southern latitudes, summer days can be so long that the sun never actually sets. Like other regions, the seasons at the North and South Poles are different depending on which hemisphere you are in. When it is summer at one pole, it is winter at the other.

CCC Patterns

Is the polar climate a type of desert? Explain why or why not, using evidence from your reading.

____

____

____

____

____

____

Think about what you have read as you **watch** the videos that follow.

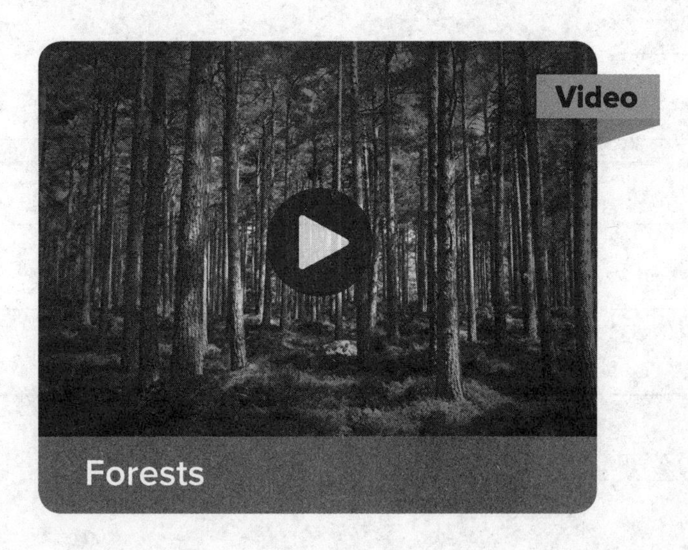

How does climate affect what kinds of trees will be found in the forest in a particular region?

How does the temperature around the equator support the tropical environment?

Plan Your Research

Choose one of the three climate regions you have learned about. **Use** the Internet and other resources to do more research on that region. You can **plan** your research and **record** your notes below.

My Climate Region: ______________________________

What I Know	What I Want to Know	What I Have Learned

Tweet, Tweet

Choose a day sometime in the next year. What would the weather be like on that day in each of the three regions you have learned about? **Write** a social media post giving the weather forecast for each location on the date you chose.

Some details you may want to include are

- location
- precipitation
- temperature
- suggested outdoor activities

Draft your social media post below.

Activity 12

Analyze Like a Scientist

Quick Code:
us3770s

Predicting Rain

Read the text. Then, **answer** the questions.

Predicting Rain

The climate of a region helps us **predict** weather conditions. If you went to the desert right now, do you think you would see rain? Do you think the Arctic is warm enough to go for a swim in the ocean? We know that the desert does not get rain often, and the Arctic is a polar region, so it should be very cold at the North Pole right now.

Forecasters use the history of a region to predict what the weather will be like in the future. They find patterns in the climate to support their predictions. If the climate in a region shows a history of rain every April, a forecaster will predict rain next April. If you live in a region that has snow every winter, you can predict that you will need warm clothes next winter.

Why is it important to predict the weather?

____

____

____

____

How does the average rainfall in your community compare to the other regions you have studied?

____

____

____

____

Activity 13

Evaluate Like a Scientist

Quick Code:
us3771s

Predicting Regional Climate

Look at the pictures. Which areas would get the most rain? The least? Which areas would have the highest temperatures? The lowest? **Draw** a line from each picture to the temperature and amount of precipitation that are typical for that region.

Coastal

Desert

Polar

Tropical

84°F, 4 inches

64°F, 4 inches

47°F, 0 inches

–37°F, 0.23 inches

Activity 14

Record Evidence Like a Scientist

Quick Code:
us3772s

Droughts

Now that you have learned about regional climates, look again at the Droughts map. You first saw this in Wonder.

Let's Investigate Droughts

Talk Together

How can you describe droughts now? How is your explanation different from before?

SEP **Constructing Explanations and Designing Solutions**

Look at the Can You Explain? question. You first read this question at the beginning of the lesson.

Can You Explain?

How does knowing the climate of different regions help predict the type of weather that a region generally experiences?

Now, you will use your new ideas about droughts to answer a question.

1. **Choose** a question. You can use the Can You Explain? question or one of your own. You can also use one of the questions that you wrote at the beginning of the lesson.

2. Then, use the graphic organizers on the next pages to help you answer the question.

How does knowing the climate of different regions help predict the type of weather that a region generally experiences?

To plan your scientific explanation, first **write** your claim.

My claim:

Next, **look** at your notes in the Bubble Map. Identify two pieces of evidence that support your claim:

Evidence 1

Evidence 2

Bubble Map

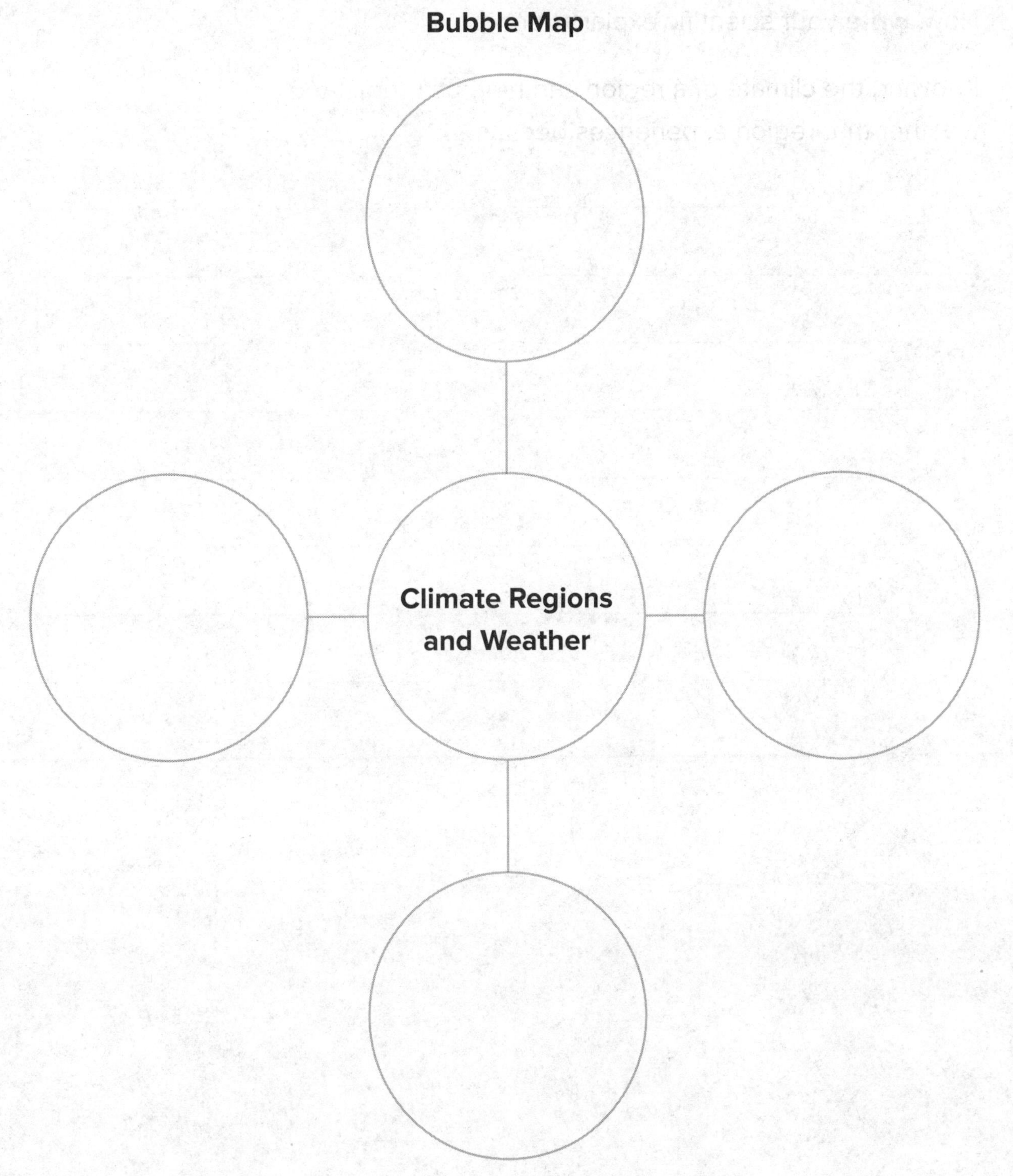

How does knowing the climate of different regions help predict the type of weather that a region generally experiences?

Now, **write** your scientific explanation.

Knowing the climate of a region can help us predict the weather that region experiences because...

STEM in Action

Quick Code: us3773s

Activity 15

Analyze Like a Scientist

Read the text about climatologists. Then, **complete** the activity that follows.

Climatologists

How do you know about things you cannot see? What if you cannot hear them? Or touch them? What if they happened before you were even born? Climatologists investigate the climate, and often the climate from thousands of years ago. So, how do they do it? Climatologists need to find ways to investigate the **atmosphere** from many years ago.

Climatologists

SEP **Obtaining, Evaluating, and Communicating Information**

Climatologists *cont'd*

In addition to ice cores and tree rings, climatologists run computer simulations. These simulations can predict general trends. The latest simulations show that the climate is warming.

There is other information to support this conclusion. Ice cores allow climatologists to study the atmosphere from many years ago. The ice cores contain little bubbles of air from long ago. These ice cores show that greenhouse gases have been increasing in the atmosphere.

Ice Core Investigations

Imagine you could examine an ice core in your town or region over the past 50 years. Do some **research** on the climate in your area over the past year, and create a scientific journal detailing the air quality, seasonal weather patterns, and other climatic events that took place. **Compare** that information to what Earth's climate was like 50 years ago in your area.

Climate over the Past Year	Climate 50 Years Ago

Activity 16

Evaluate Like a Scientist

Quick Code: us3774s

Review: Regional Climates

Think about what you have read and seen. What did you learn?

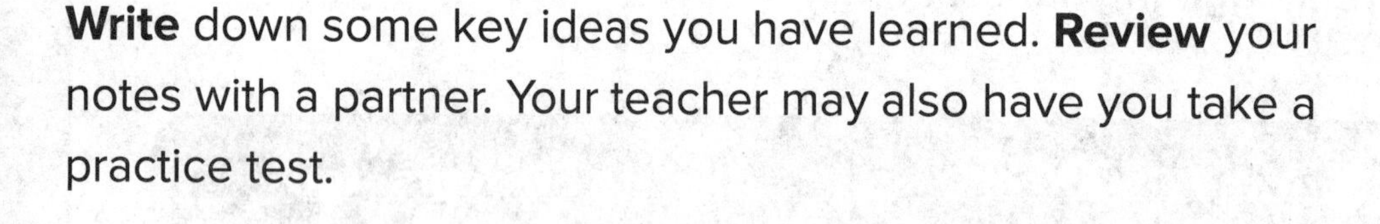

Write down some key ideas you have learned. **Review** your notes with a partner. Your teacher may also have you take a practice test.

SEP **Obtaining, Evaluating, and Communicating Information**

Talk Together

Think about what you saw in Get Started. Use your new ideas to discuss why different regions of the world have different climates.

CONCEPT

4.2

Predicting Weather

Student Objectives

By the end of this lesson:

- [ ] I can determine the best methods and tools for collecting weather data in different areas.
- [ ] I can explain how meteorologists use data collected from many sources to describe weather patterns that can be used to predict future weather.
- [ ] I can use weather data to create tables and graphical displays that show weather patterns and relationships.
- [ ] I can analyze and interpret patterns in weather data to predict future weather conditions.

Key Vocabulary

- [ ] air pressure
- [ ] barometer
- [ ] data
- [ ] detect
- [ ] meteorology
- [ ] observe
- [ ] wind

Quick Code:
us3776s

Activity 1

Can You Explain?

How is weather data gathered and used to forecast the weather?

Quick Code:
us3777s

Activity 2

Ask Questions Like a Scientist

Quick Code:
us3778s

Hurricane Wilma

Look at the satellite image of Hurricane Wilma. What do you observe?

Let's Investigate Hurricane Wilma

SEP **Using Mathematics and Computational Thinking**

What do you think the weather will be like for the next few days in this area? **Write** your prediction and **share** it with a partner.

What do you wonder about making a weather forecast? **Write** three questions you have and **share** them with the class.

I wonder...

I wonder...

I wonder...

Activity 3

Analyze Like a Scientist

Quick Code:
us3779s

What Is Weather?

Read the text and **look** at the picture of Hurricane Wilma. When you reach a question, **write** your answer to the question in the box below. **Share** and **discuss** your answer with a partner. Do this with every question in the text.

What Is Weather?

Do you recognize the weather feature shown in the picture? What is happening with the cloud formations in the sky? What do you imagine the weather is like on the ground? If you could measure the temperature inside the storm, what do you predict it would be? Would the temperature be changing from one minute to the next? What would the **wind** speed be, and would it be changing?

What will the weather be like tomorrow, or the day after? Forecasters predict the weather. In this concept, you will learn why predicting the weather is important and how forecasters make their predictions.

Hurricane Wilma

Being able to make observations and predictions about the weather are valuable skills. They can help you make important decisions such as what clothes to wear, what activities to plan, and when it is time to run for cover!

My Answers

Activity 4

Observe Like a Scientist

Quick Code: us3780s

What Is Weather Forecasting?

Watch the videos. **Look** for details about weather forecasting that you can use to fill in the graphic organizer. **Record** the details you find in the correct spaces of the graphic organizer.

Video

Today's Weather

Video

The Importance of Weather Forecasting

Video

The Weather Is Different from Day to Day and from Place to Place

Definition - **Personal** - **Dictionary**	**Examples** (drawn or written)
Sentences - **Teacher/Book:** - **Personal**	**Related Word** Parts:

Weather Forecasting

Outside of School (Who would use the word? How would they use it?):

Activity 5

Evaluate Like a Scientist

Quick Code:
us3781s

What Do You Already Know About Predicting Weather?

Weather Data Instruments

Look at the images of the different instruments used to collect data on weather. **Draw** lines connecting the images with their correct names. It is okay if you do not know all of the answers yet.

Anemometer

Barometer

Rain Gauge

Thermometer

Weather Forecasting

Which statement about weather forecasting is correct? **Circle** the answer.

A) With modern technology, weather forecasting is extremely accurate.

B) If a meteorologist can accurately read a weather map, he or she can accurately forecast the weather.

C) Because weather is constantly changing, weather forecasting is a fairly uncertain science.

D) Most weather can be accurately predicted with a thermometer, barometer, and wind gauge.

How Do People Use Common Tools to Measure Weather Conditions?

Activity 6

Investigate Like a Scientist

Quick Code: us3782s

Hands-On Investigation: Collecting Weather Data

In this activity, you will set up a weather data station. Every day, you will collect information using a wind vane, rain gauge, compass, clock, camera, and Celsius thermometer. You will observe, measure, and record the weather at the same time every day. You will compare your weather data with the daily forecast. You will compare your data with data from a group in a different location.

Make a Prediction

Question	Prediction
How accurate do you think the daily forecast will be?	
Do you think the data you collect will be the same as or different from the other groups' data?	

SEP **Asking Questions and Defining Problems**

CCC **Patterns**

What materials do you need? (per group)

- Wind vane
- Rain gauge
- Thermometer, plastic
- Camera
- Compass
- Clock

What Will You Do?

Preparation

1. If you have not already constructed a wind vane, do so now.
2. Construct a rain gauge by placing a plastic ruler against the outside of a jar. Line up the "0" line on the ruler with the *inside* bottom of the jar. Use an elastic band to hold the ruler in place. Set a funnel on top of the jar.
3. Place your rain gauge on a level surface outside, away from trees and sprinklers. You can surround it with pebbles to keep it from tipping over.
4. Using your Weather Data Sheet, record the location, date, current weather conditions, and measurements from the different instruments.

Daily Recording

1. You will be recording the forecasts and actual weather conditions each day for 14 days.
2. Record the weather forecast for each day either the day before or early in the morning.
3. In your group, return to the same location at the same time every day to record the conditions on your Weather Data Sheet. Write in one color for the first seven days and write in a different color for the next seven days. Make sure to empty the rain gauge after each daily recording.

Graphing

Once your data have been collected, create a line graph of the daily temperatures and a bar graph of the daily precipitation.

Weather Data Sheet

Name	
Date	
Location	

Weather variable:	Observations:	Measurements:	Time of measurements:
Temperature (°C)			
Wind			
Precipitation			
What the sky looks like			

Graph for Temperature

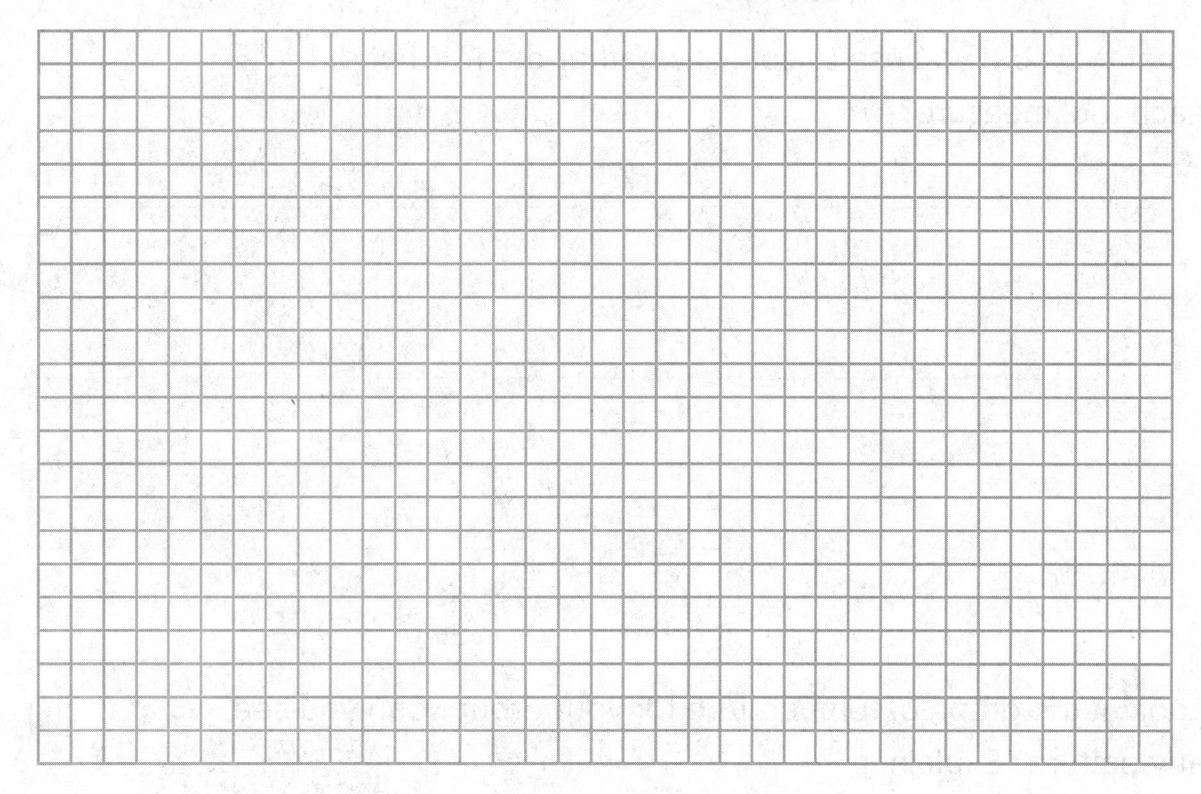

Graph for Precipitation

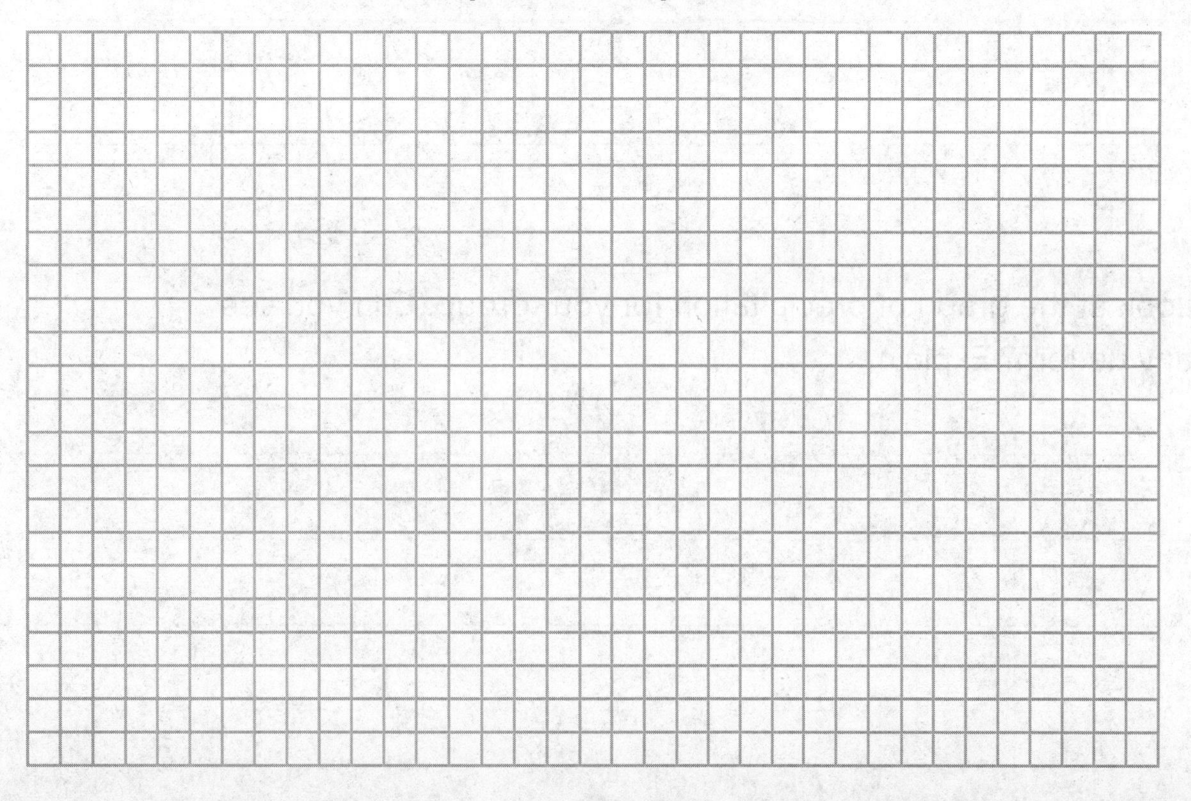

Think About the Activity

What tools did you use to collect weather data? What did each tool measure?

Look at the graph of temperature for your group. Can you see any pattern? Explain.

Look at the graph of precipitation for your group. Can you see any pattern? Explain.

Choose one day. Compare your weather data with the weather forecast.

Choose two days. Compare your weather data for these two days.

Work with another group. Compare your graphs. If you see something different, try to explain why it might be different.

Activity 7

Observe Like a Scientist

Quick Code:
us3783s

Weather Tools

Watch the video and **record** how some of the tools can be used to describe the weather. **Draw** some of the tools in the boxes of your graphic organizer, and **write** information about them on the lines below each box. **Discuss** what you learned from the video with a classmate.

Weather Smart: Forecasting and Weather Instruments

Talk Together

Now, talk together about the different tools shown in the video. What do they do? How do they work?

Use this graphic organizer to record what you are learning about weather tools.

Drawing of Tool	Drawing of Tool	Drawing of Tool

Description of Tools

Drawing of Tool	Drawing of Tool	Drawing of Tool

Description of Tools

Activity 8

Analyze Like a Scientist

Quick Code:
us3784s

How Much?

Read the texts and **record** any new information in your graphic organizer from the first activity in this section. **Draw** some of the tools in the boxes of your graphic organizer, and **write** information about them on the lines below each box. You may add new information from this activity to tools you have already drawn and written about. **Compare** the photos in the texts to objects you have seen before.

How Much?

There are two ways to describe the weather. One way is to describe what you see. For example, you can say that it is sunny or that it is windy. Another way is to measure. For example, you can say that the temperature is 22°C (72°F) or that the wind is blowing at 20 kilometers per hour.

Notice that measuring gives you specific information about the weather. You know exactly how warm and windy it is outside. Measuring answers the question: *How much?* Scientists use different instruments to collect information about the weather.

Here, you can see a thermometer, a barometer, and other instruments for measuring the weather.

A thermometer measures temperature. A German scientist invented one of the first thermometers that had numbers on it. His name was Daniel Fahrenheit. However, scientists do not use this type of thermometer. They use a temperature scale that was developed by a Swedish scientist named Anders Celsius. On the Celsius scale, water freezes at 0° and boils at 100°. The highest outside temperature people have ever measured is 57.7°C (136°F). The weather became this hot in North Africa one afternoon in 1922.

An anemometer measures wind speed. In 1450, an Italian invented the first anemometer. The world record for highest wind speed was recorded in April 1934 on Mount Washington, New Hampshire.

This anemometer measures the speed of the wind.

A **barometer** is a device that measures **air pressure**. Around 1640, an Italian named Evangelista Torricelli invented the first barometer. A drop in air pressure can mean a bad storm, such as a hurricane, is approaching. The lowest air pressure in a hurricane that went on to strike the United States was measured in 2005.

A rain gauge measures precipitation. The most common rain gauge used today was invented over a hundred years ago. The rain gauge is simply a cup. Inside the cup is a funnel that drains into a tube. Precipitation, such as rain and snow, falls inside the funnel. The water then collects in the tube. A ruler can then be used to measure how much precipitation fell. Are you ready for another world record? The most rain that fell in one day soaked an island off the coast of Africa in 1980. Over 115 centimeters of rain fell that day. That is more rain than most places receive in a whole year!

How Much? Part 2

To predict and describe the weather, meteorologists use many tools, large and small.

Some small tools are thermometers, rain gauges, anemometers, and barometers. A thermometer measures temperature. Temperature is the warmth or coldness of the air. A rain gauge shows how much precipitation has fallen. Scientists keep this gauge outside. When the rain stops, they measure the water in the gauge. An anemometer shows how fast the wind blows. This is the wind speed. A barometer shows the air pressure. Air pressure tells scientists the likelihood of storms. High air pressure makes storms less likely. Low air pressure makes it easier for storms to form.

Some large tools are satellite and radar. The National Weather Service uses these instruments to create weather data. The data is another tool. Satellites help meteorologists study weather from space.

Satellite images and data are two resources meteorologists use.

How Much? Part 2 *cont'd*

The satellites give meteorologists images and information. These data can show temperature, humidity, and wind. Radar is useful to learn about large-scale events. These include thunderstorms and hurricanes. Doppler is a special form of radar that meteorologists use. The National Weather Service gets information from almost 1,000 weather stations around the United States. The data it provides is a useful tool that helps scientists study and predict weather patterns.

Activity 9

Observe Like a Scientist

Quick Code:
us3785s

Weather Data

Complete the interactive to learn more about different tools used in weather forecasting. **Record** your findings in your graphic organizer from the first activity in this section. **Draw** some of the tools in the boxes of your graphic organizer, and **write** information about them on the lines below each box. You may add new information from this activity to tools you have already drawn and written about.

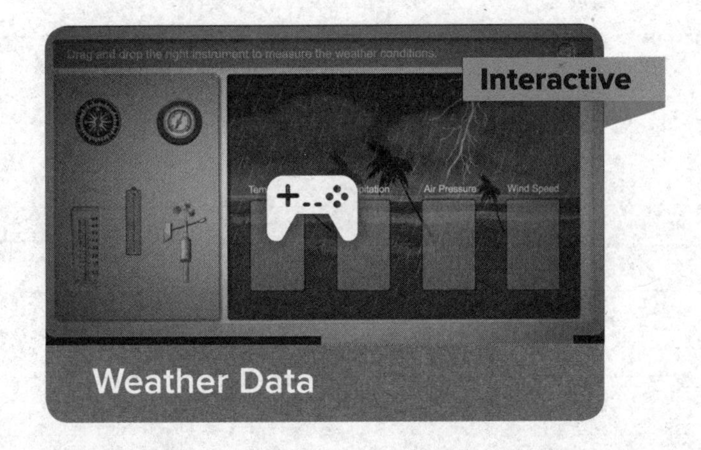

Activity 10

Evaluate Like a Scientist

Quick Code:
us3786s

Using Weather Tools

Write three truths and one lie about weather tools. Then, **share** your statements with a classmate.

SEP **Analyzing and Interpreting Data**

Read the reports describing weather in different places. **Determine** which weather tools would be needed to make each report, then **draw** connecting lines between the reports and the tools needed. More than one tool may be appropriate for each report.

Reports
Sunny skies are likely because the air pressure is rising.
Button up your coat, because the wind is blowing at 12 miles per hour.
Nearly an inch of water has fallen here today. As the air becomes colder, that water will freeze and we could get snow.
It is very hot outside: 101°F There is very little moisture in the air, however, so do not expect rain.

Tools
Anemometer
Barometer
Hygrometer
Thermometer
Rain gauge

How Do Meteorologists Look for Weather Patterns and Predict Weather?

Activity 11

Observe Like a Scientist

Quick Code: us3787s

Predicting Weather Patterns

Watch the video, then **discuss** what you learned from the video with a classmate.

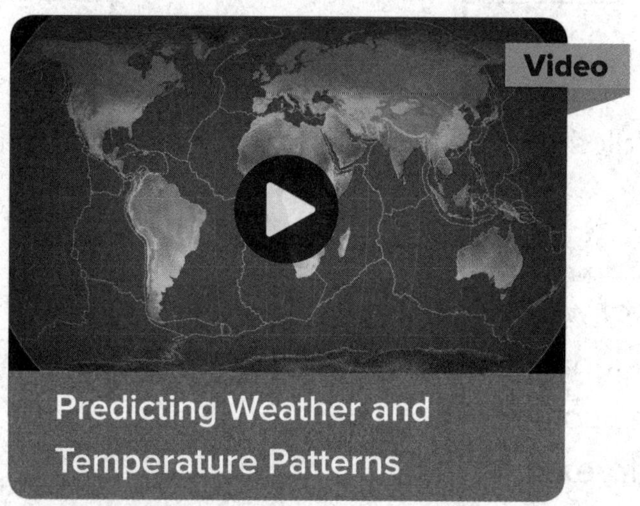

Predicting Weather and Temperature Patterns

Talk Together

What are some of the patterns commonly found in the weather? How are scientists able to record patterns and use them to make predictions?

CCC **Patterns**

Activity 12

Analyze Like a Scientist

Quick Code: us3788s

Radar and Satellite

Read the text and **watch** the video to learn more about the tools meteorologists use. You will be assigned to research a specific tool that is mentioned in the text and video. **Look** for pictures, a summary of how the tool helps predict the weather, and where the tool is often used. **Record** your findings in the space provided.

Radar and Satellite

Predicting weather is important for planning many human activities, such as when a farmer should plant seeds or when it is safe to fly an airplane. Predicting weather begins by collecting **data** to **observe** weather patterns.

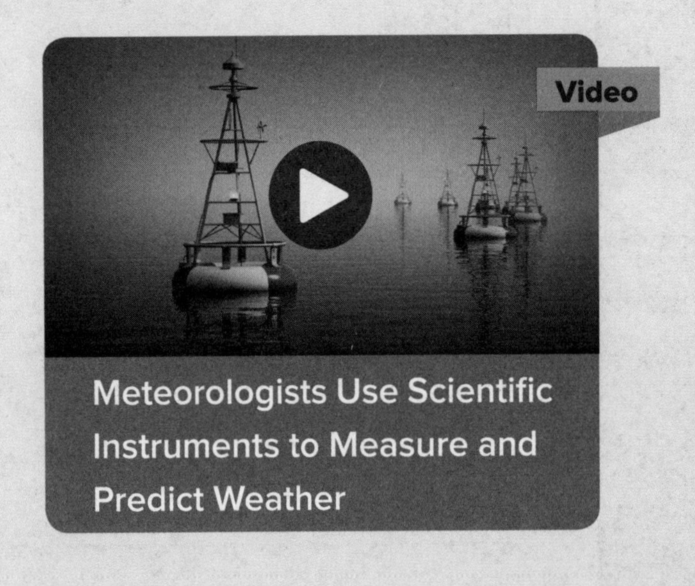

Meteorologists Use Scientific Instruments to Measure and Predict Weather

CCC **Patterns**

Radar and Satellite *cont'd*

You learned that instruments like thermometers and barometers are used to measure and record weather. Meteorologists also use a scientific instrument called radar to **detect** precipitation. Radar helps track thunderstorms and hurricanes. Satellites high above Earth are used to collect weather data over large areas.

My Notes:

What Are Ways We Can Present Weather Data?

Activity 13

Observe Like a Scientist

Quick Code:
us3789s

Organizing a Forecast

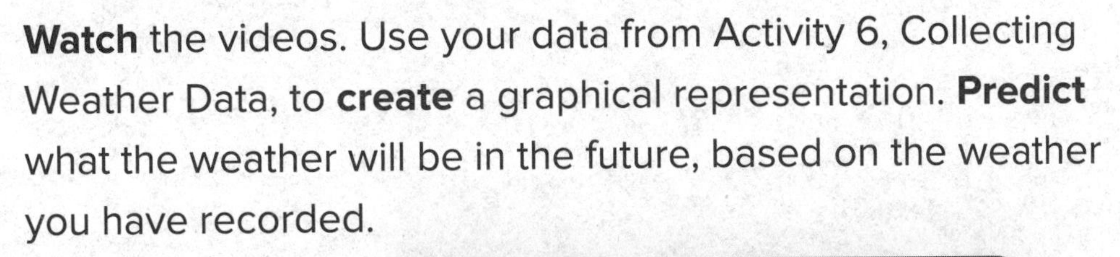

Watch the videos. Use your data from Activity 6, Collecting Weather Data, to **create** a graphical representation. **Predict** what the weather will be in the future, based on the weather you have recorded.

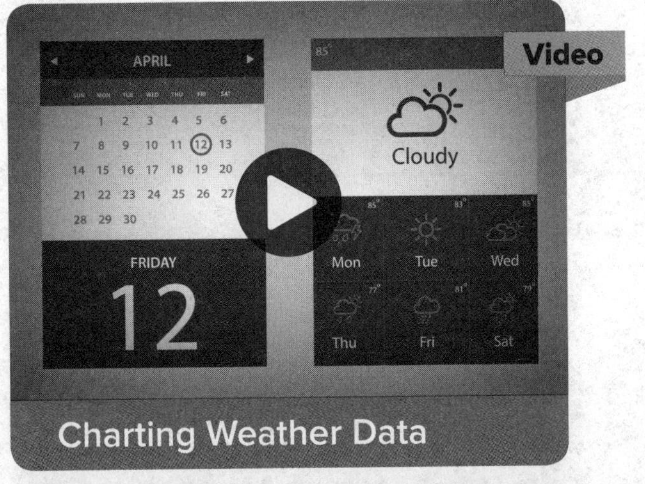

Charting Weather Data

Organizing Data

Draw your graphical representation of the data you collected in the space provided.

Predict the upcoming weather. Use the Claim-Evidence-Reasoning Chart to explain your thinking.

Claim	
Evidence	
Reasoning	

SEP **Analyzing and Interpreting Data**

Activity 14

Evaluate Like a Scientist

Quick Code:
us3790s

Predicting the Weather from Patterns

California in February

Look at the weather patterns shown in the calendar for California in February. Based on the data, **answer** the questions.

Su	M	T	W	Th	F	S
1	2	3	4	5	6	7
8	9	10	11	12	13	14
15	16	17	18	19	20	21
22	23	24	25	26	27	28

SEP Analyzing and Interpreting Data

What do you predict the weather will be like next February in California?

Explain your prediction.

Average Temperature

Look at the bar graph of temperatures in northern California throughout the seasons. Based on the data, **answer** the questions.

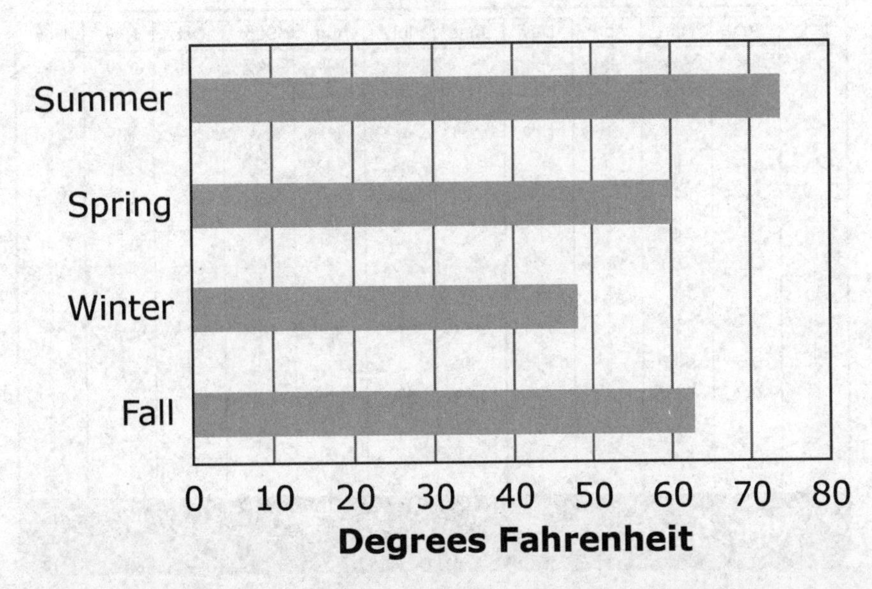

What do you predict the average temperature will be in northern California this summer?

Explain your prediction.

Wind Speeds

Look at the table comparing average wind speed throughout the year at a single point in Sacramento. Each image of an anemometer indicates 1 mile per hour (mph). Based on the data, **answer** the questions.

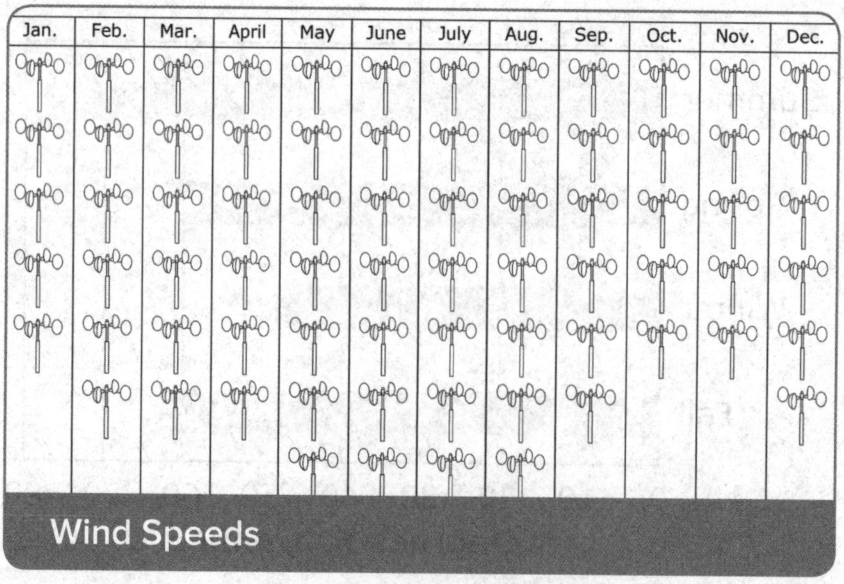

Jan.	Feb.	Mar.	April	May	June	July	Aug.	Sep.	Oct.	Nov.	Dec.

Wind Speeds

What do you predict the average wind speed will be in Sacramento next June?

Explain your prediction.

Activity 15

Record Evidence Like a Scientist

Quick Code:
us3791s

Hurricane Wilma

Now that you have learned about gathering weather data to forecast weather, look again at the image of Hurricane Wilma. You first saw this in Wonder.

Let's Investigate Hurricane Wilma

Talk Together

What would your weather prediction for the next few days in this area be now?

How is your explanation different from before?

SEP **Constructing Explanations and Designing Solutions**

Look at the Can You Explain? question. You first read this question at the beginning of the lesson.

Can You Explain?

How is weather data gathered and used to forecast the weather?

Now, you will use your new ideas about making a forecast to answer a question.

1. Choose a question. You can use the Can You Explain? question or one of your own. You can also use one of the questions that you wrote at the beginning of the lesson.

2. Then, use the graphic organizer on the next page to help you answer the question.

To plan your scientific explanation, first **write** your claim.

My claim:

Next, **look** at your notebook or journal. Identify two pieces of evidence that support your claim (specifically evidence that shows patterns):

Evidence 1

Evidence 2

Now, **write** your scientific explanation.

STEM in Action

Quick Code:
us3792s

Activity 16

Analyze Like a Scientist

Making Weather Information Available to the Public

The National Weather Service (NWS) is a large governmental organization. Many of the people who work for the NWS are weather scientists. They are responsible for providing U.S. weather data to the public. They create detailed weather maps of the United States. Local weather forecasters use this information to create their forecasts. They also use it to create the maps you see on the local news. This reading passage describes some of the work of the NWS. **Read** the passage, and then **complete** the activity that follows.

SEP **Obtaining, Evaluating, and Communicating Information**

Now, Here's Your Local Weather

"Now, here's your local weather." When you hear these words on television, you can be sure that you will see a weather map. A weather map shows weather patterns. A weather map also provides information that is used to forecast the weather. If you look at a series of weather maps, you can see how conditions like wind direction, temperature, and precipitation change from day to day.

The National Weather Service (NWS) collects information from about 1,000 weather stations across the United States. The NWS uses this information to make a weather map of the entire country. This map uses small, key-shaped symbols. Each symbol shows the location of a weather station. The symbols tell how cloudy it is. If the circle of the key is filled in, it means the sky is cloudy at that weather station. The long arm of the key shows the wind direction. For example, if the arm of the key points up (north), the wind is coming from the north. The short lines on the end of the key tell how strong the wind is. The symbols also show weather conditions such as rain and snow.

The NWS map has more details than a weather map on television. A television weather map usually shows areas of low pressure and high pressure.

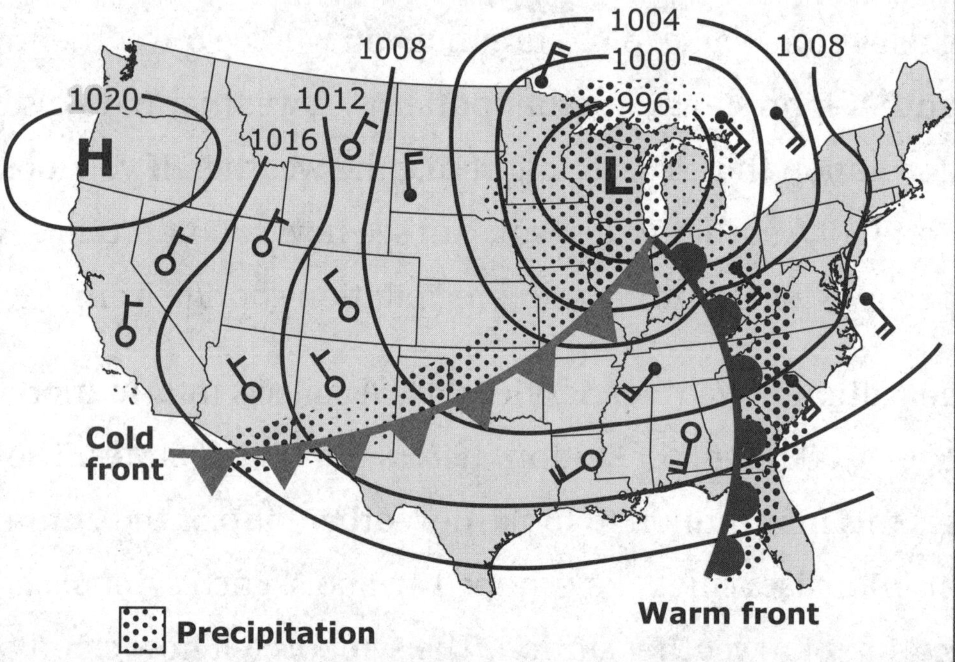

Fronts and Precipitation

Weather maps can give you much information about the weather.

A capital L represents an area of low pressure. A capital H represents an area of high pressure. A television weather map also usually shows weather fronts. For example, a weather map may show a cold front approaching your area. This means that cold air is pushing out warm air. The cold front is shown on a weather map as a blue curved line with small triangles. The triangles point in the direction the front is heading.

The information on a weather map is useful in forecasting the weather. If a high-pressure area is approaching you, then you will likely have fair weather. If a warm front is approaching, then you might have some rain. Warmer, more **humid** weather will follow. Weather forecasting is much easier if you know how to read a weather map.

Predicting the weather can be very difficult. There are many factors that affect weather. And these factors can change at any minute! Having as much data as possible helps weather scientists and forecasters do their job.

Using the National Weather Service

Maps taken from the National Weather Service website show the weather conditions at the date and time shown at the top of the map.

You have learned that NWS scientists use the data they collect to provide information to other meteorologists and to issue weather warnings. But what information can the NWS website provide to you? **Explore** the NWS website. **Find** some maps that show weather conditions. Then, **answer** these questions:

Can you find your local weather forecast? If so, how?
Hint: Find out what your zip code is!

Are there currently any weather alerts or warnings for your area? How do you know?

Return to the NWS home page and click on the area where you live. The page that appears shows local weather conditions and also names, near the top of the page, your local NWS Forecast Office. What is the location of the NWS office closest to you?

Activity 17

Evaluate Like a Scientist

Quick Code: us3793s

Review: Predicting Weather

Think about what you have read and seen. What did you learn?

Write down some key ideas you have learned. **Review** your notes with a partner. Your teacher may also have you take a practice test.

SEP **Obtaining, Evaluating, and Communicating Information**

Talk Together

Think about what you saw in Get Started. Use your new ideas to discuss how weather can be predicted.

CONCEPT
4.3

Weather Hazards

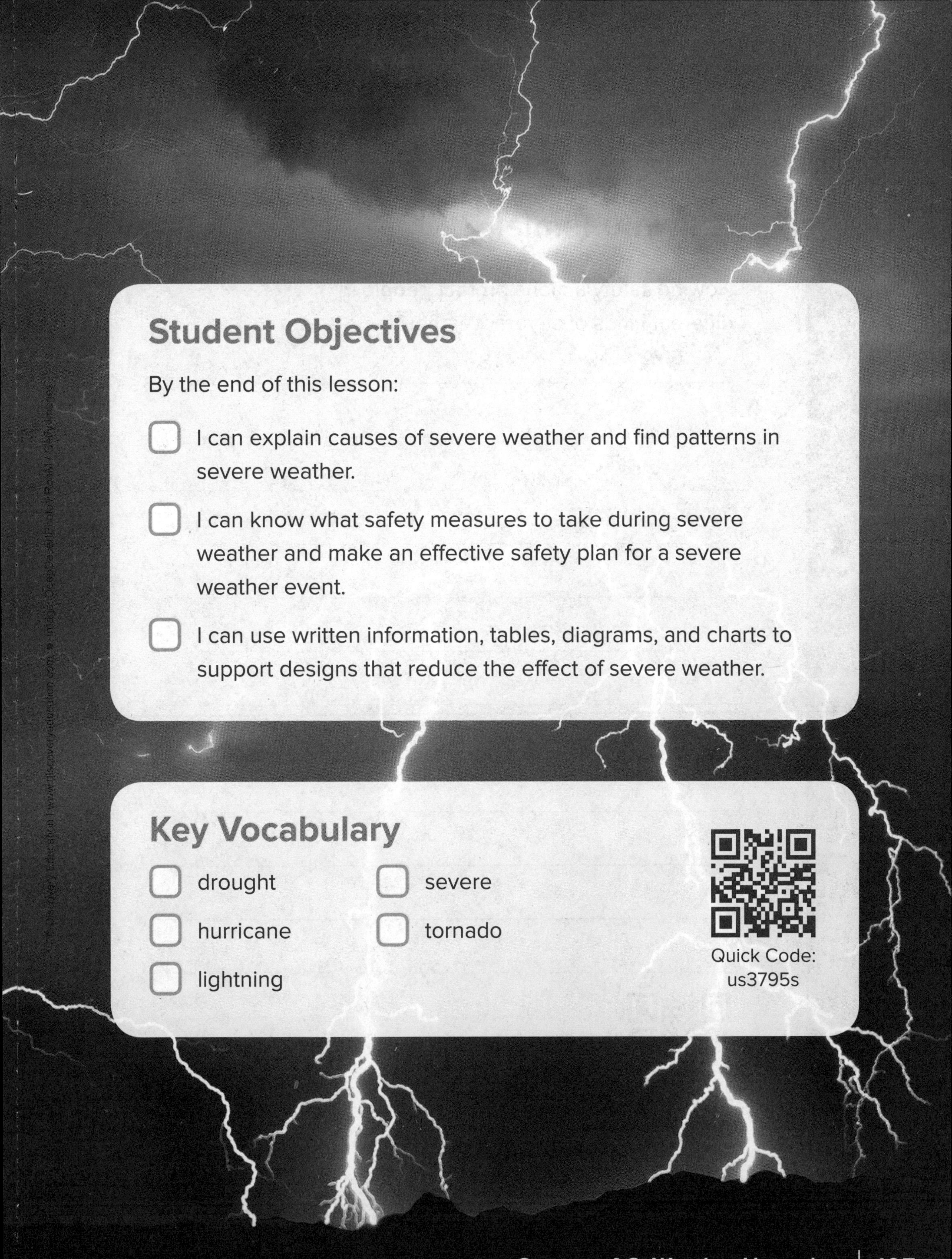

Student Objectives

By the end of this lesson:

- ☐ I can explain causes of severe weather and find patterns in severe weather.
- ☐ I can know what safety measures to take during severe weather and make an effective safety plan for a severe weather event.
- ☐ I can use written information, tables, diagrams, and charts to support designs that reduce the effect of severe weather.

Key Vocabulary

- ☐ drought
- ☐ hurricane
- ☐ lightning
- ☐ severe
- ☐ tornado

Quick Code: us3795s

Activity 1

Can You Explain?

How do safety actions protect people in different kinds of severe weather?

Quick Code:
us3796s

Activity 2

Ask Questions Like a Scientist

Quick Code:
us3797s

Mudslide

Look at the image of a mudslide. **Answer** the questions in the space provided.

Let's Investigate Mudslide

What could have caused the mudslide?

What could have been done to prevent the damage of the mudslide?

Activity 3

Analyze Like a Scientist

Quick Code:
us3798s

What Is Severe Weather?

Read the text about severe weather. **Fill in** the What I Know and What I Want to Know sections of the KWL Chart. You will fill in the What I Learned section as you work through the concept.

What Is Severe Weather?

Have you ever seen or heard a weather report? Most of the time, we check the weather to find out what kind of clothes to wear or whether we will need an umbrella. Sometimes, though, the weather report warns that severe weather is coming. What are some examples of severe weather? What severe weather is common in your community?

What other kinds of severe weather happen around the world?

In this concept, you will learn about different types of severe weather. You will also learn ways to prepare for severe weather that can help you stay safe.

Severe Weather: ______________________________

What I Know	What I Want to Know	What I Have Learned

Activity 4

Observe Like a Scientist

Quick Code:
us3799s

What Is Severe Weather?

Watch the videos and **look** at the image of different types of severe weather. **Record** any questions you have about the different types of severe weather on the Question side of the T-Chart provided. **Write** any connections you have to each type of weather (Has it happened in your community? Have you experienced it in your life?) on the Connections side.

Hail

Tornado

Blizzard

Severe Thunderstorms

How do safety actions protect people in different kinds of severe weather?

Severe Weather Examples: ______________________________

Questions	Connections

Activity 5

Evaluate Like a Scientist

Quick Code: us3800s

What Do You Already Know About Weather Hazards?

Severe Weather Warning

A warning is sent out by a local weather forecaster that severe winter weather is coming to the area. There will be strong winds, causing blowing snow. It will be difficult to see.

Circle the kind of weather that is most likely coming to the area.

A) Blizzard

B) Thunderstorm

C) Hurricane

D) Tornado

Safety Kit Materials

Some materials for a family storm emergency kit are listed below. **Number** the materials from most (1) to least (5) important. Think about them in terms of staying safe at home during the severe weather.

___ Magazines and books

___ Cash

___ First aid kit and necessary medications

___ Flashlight

___ Extra batteries

Prepping for Severe Weather

Suppose you find out that a hurricane will be coming to your area by tomorrow. **Write down** at least two things you can do to prepare your home and/or remain safe during the storm and **explain** why these things should be done.

____

____

____

____

____

____

____

____

Severe Weather Facts

Circle the statements about severe weather and severe weather safety that are correct.

A) It is safe to drive on flooded roads.

B) Lightning never strikes twice in the same place.

C) There is nothing we can do to prepare for severe weather.

D) Hurricanes and tornadoes are different kinds of severe weather.

E) Preparing for severe weather can help you stay safe.

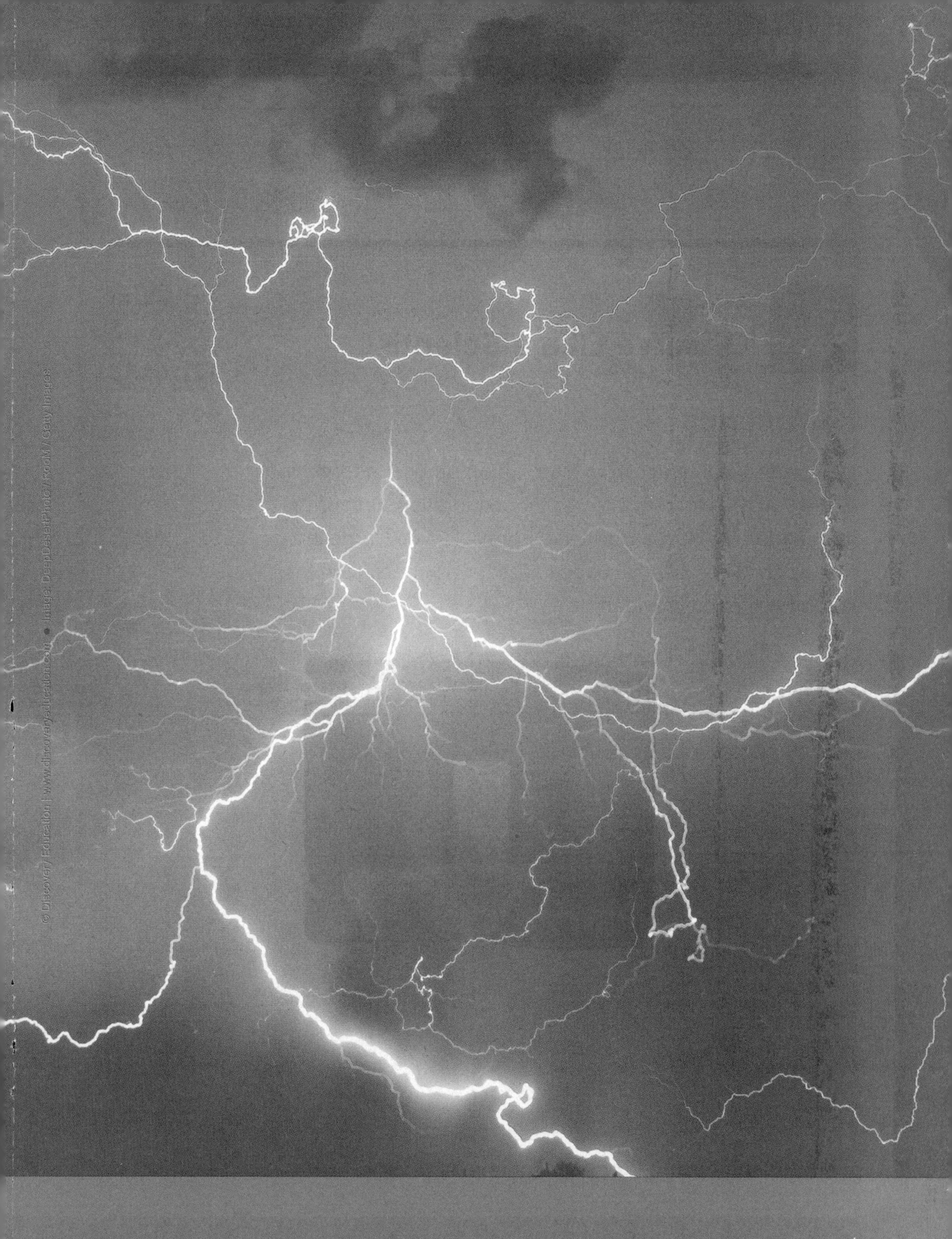

What Causes Different Types of Severe Weather?

Activity 6

Observe Like a Scientist

Quick Code:
us3801s

Hurricanes

Complete the interactive to learn about hurricane formation. While you are working through the exploration, **record** the information for How a Hurricane Forms, How a Hurricane Moves and Dies, and Damage Left after a Hurricane in the summary frames provided.

Hurricanes

SEP Planning and Carrying Out Investigations

How a Hurricane Forms	How a Hurricane Moves and Dies	Damage Left after a Hurricane

Activity 7

Analyze Like a Scientist

Quick Code: us3802s

Severe Weather

Read the text. **Circle** the types of severe weather, and **underline** evidence for why they are severe.

Severe Weather

What makes some weather extreme?

We often drive or walk through ordinary wind, rain, or snow. But **severe** weather is more dangerous. Thunderstorms come with heavy rain and dangerous **lightning**. Hurricanes produce strong winds and rain that can lead to floods. Tornadoes are swirling, twisting winds so strong they can lift cars and destroy houses. Blizzards are snowstorms that make it hard to see, drive, or even walk safely.

A **drought** occurs when an area has little or no rain for a long period of time. Forests that experience drought conditions are at increased risk of wildfires.

Activity 8

Observe Like a Scientist

Quick Code:
us3803s

Tornadoes and Hurricanes

Watch the videos. As you watch, **fill in** the Venn diagram comparing tornadoes and hurricanes. Note how each storm is formed, where each storm is formed, where each storm causes damage, and what type of damage each storm causes.

How Tornadoes Develop

How Hurricanes Develop

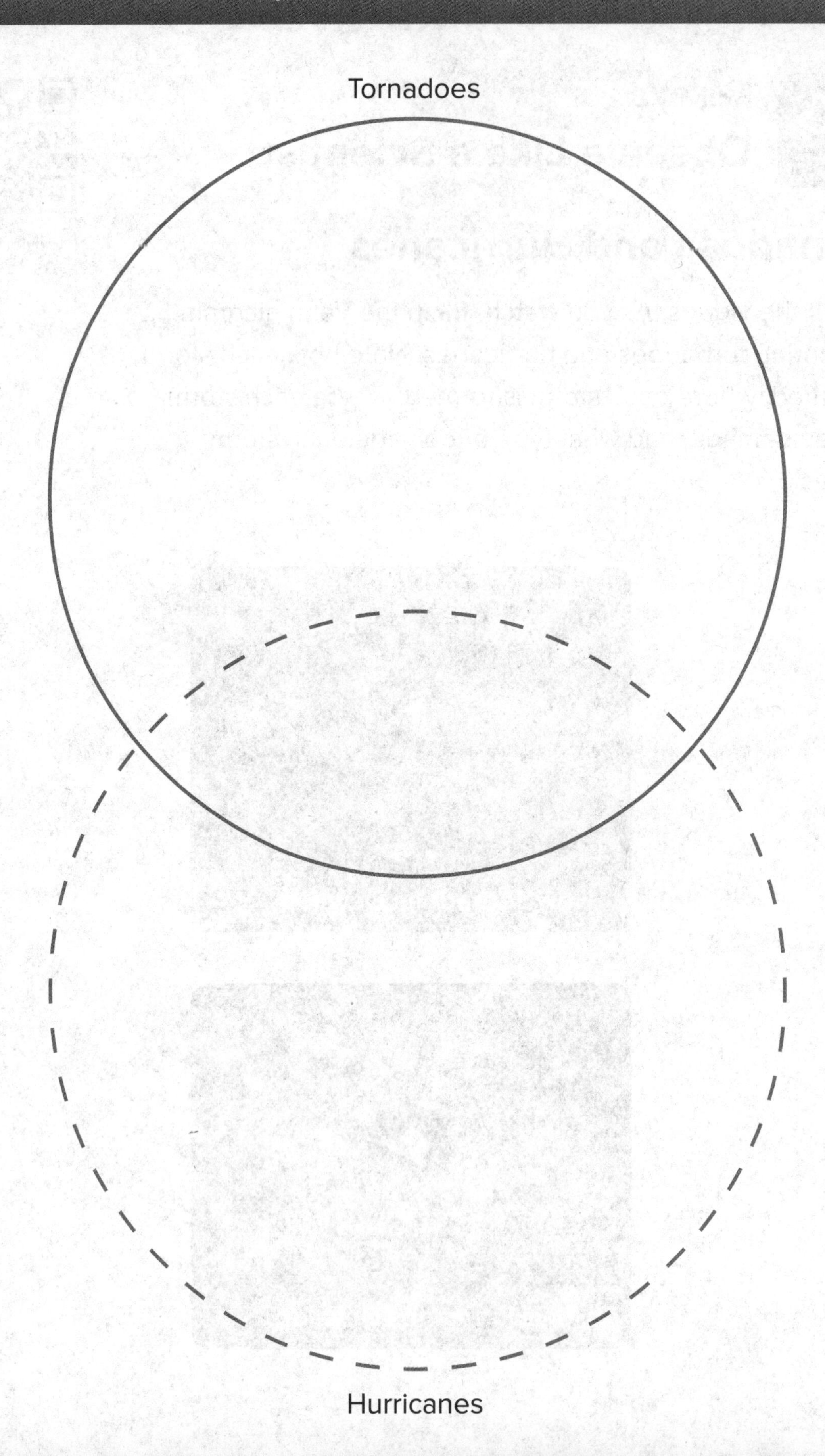
Tornadoes
Hurricanes

Activity 9

Observe Like a Scientist

Quick Code:
us3804s

Droughts and Floods

You will be assigned a topic for a classroom debate. Your task is to argue that your topic is the more dangerous of the two. If your topic is droughts, **read** the text and **watch** the videos. If your topic is floods, **read** the text and **watch** the video. **Record** facts about the dangers of your topic that you will use in your argument in the area provided.

Droughts and Floods

Droughts

Droughts can occur over a short period (a few weeks) or a long period (a few years) and can happen anywhere in the world. Droughts happen when a place receives less than normal amounts of rain or snow during a season.

Wildfires in the Northwest

Drought

Floods

Floods occur when rainfall in an area is greater than the soil can absorb and greater than the rivers can hold. Floods are most common in spring after multiple days of rain combined with melting snow.

Floods

My claim:

Evidence I found:

My claim is true because:

Activity 10

Analyze Like a Scientist

Quick Code: us3805s

Types of Severe Weather

Types of Severe Weather

Blizzard

Each type of severe weather occurs for a particular reason. Some forms of severe weather only occur at certain places or times of year. Example are blizzards, hurricanes, tornadoes, droughts, and floods.

Blizzards can occur only when the weather is cold enough to freeze water (0°C). Blizzards are common in winter where the weather gets very cold.

Hurricanes form over warm oceans. Most hurricanes occur in tropical places in the late summer and early fall, when the water is warmest.

Tornadoes often form when cool, dry air collides with very warm, humid air. In the United States, this usually happens in an area called Tornado Alley. This flat region experiences many tornadoes. It includes Oklahoma, Texas, Kansas, Nebraska, and several other states.

Flood

Droughts can occur over a short period (a few weeks) or a long period (a few years) and can happen anywhere in the world. Droughts happen when a place receives lower than normal amounts of rain or snow during a season.

Floods occur when rainfall in an area is greater than the soil can absorb and greater than the rivers can hold. Floods are most common in spring after multiple days of rain combined with melting snow.

Choose a severe weather event to research that is not a tornado, hurricane, drought, or flood. Options for severe weather events to research include blizzards, severe thunderstorms, dust storms, or hail or ice storms. **Go online** to research the event. **Record** the cause, location, and common damage caused by the severe weather event you chose in the provided graphic organizer.

Event	
Cause	
Locations	
Damage Caused	

Draw a very basic Fakebook page for a person experiencing the weather event you chose in the space below. The Fakebook page only needs an image, where the person is from, and a newsfeed with status updates. You can include drawings as “photos” in the feed. The newsfeed should include updates on the weather before, during, and after the event, as well as information about what was damaged in the storm.

Activity 11

Evaluate Like a Scientist

Quick Code: us3806s

What Weather Is This?

Revisit the questions you wrote for the chart in Activity 3: What Is Severe Weather? Work as a class to **find answers** to everyone's questions. If there are any unanswered questions at the end of the discussion, **write** them in the space provided so the whole class can look for the answers as you learn more about severe weather.

SEP **Analyzing and Interpreting Data**

Draw a line to match the weather data with the weather event picture.

Weather Data
Time of year: Late summer Outside Temperature: 88°F Location: Tropical
Time of year: Winter Outside temperature: 12°F Location: Northern United States
Time of year: Summer Outside temperature: 101°F Location: Southwestern United States
Time of year: Spring Outside temperature: 40°F Location: East Coast of the United States

Weather Events

Activity 12

Analyze Like a Scientist

Quick Code: us3807s

Severe Weather's Dangers

Read the text, **look** at the image, and **watch** the videos. Pause after each section to **discuss** what you learned. **Add** to the chart you began in Activity 3 as you learn new information.

Severe Weather's Dangers

Lightning and Heavy Rainfall

Why is it dangerous to go outside to watch lightning? Lightning is a form of electricity that can strike tall objects and travel through metal. If lightning strikes a person, it can badly injure or even kill that person. To stay safe during a lightning storm, stay away from open areas and tall trees, and avoid using anything with wires or metal pipes.

CCC **Scale, Proportion, and Quantity**

In addition to lightning, storms that produce heavy rainfall over a short period of time can cause landslides. Landslides can happen when there is heavy rain or an earthquake, or when erosion weakens the earth. During a landslide, rocks and soil form a wave of dirt or mud that can knock down trees and quickly cover everything in its path.

Lightning

Hurricanes and Tornadoes

Tornadoes and hurricanes can sometimes form from severe thunderstorms. During a **hurricane**, winds can reach speeds greater than 240 kilometers per hour. During a **tornado**, winds can reach even greater speeds. At these speeds, winds can destroy buildings and uproot trees. They can fling cars more than 100 meters, like missiles. Hurricanes near land can also cause flooding. In 2005, Hurricane Katrina hit the southeastern United States.

The winds and flooding destroyed entire neighborhoods in cities such as New Orleans.

Hurricane Katrina

Wildfire

Wildfire

Lightning can cause wildfires, although 90 percent of wildfires are caused by human activity in dry conditions.

Wildfires can break out when there are hot temperatures, dry conditions, and gusty winds. In 2017, droughts in California caused hundreds of wildfires in both the southern and northern parts of the state. They forced hundreds of thousands of people to evacuate their homes, damaged thousands of buildings, and sent clouds filled with smoke and hot ash into the air.

Talk Together

Now, talk together about your experiences with any of these or other severe weather events. What did you do to stay safe?

How Can You Protect Yourself from Severe Weather?

Activity 13

Evaluate Like a Scientist

Quick Code: us3808s

What Should You Do?

The following passage describes the right way to handle a severe weather situation. One sentence is incorrect. **Highlight** the incorrect sentence.

Joseph was walking home from school when he heard thunder. A moment later, he saw a flash of lightning. Joseph quickly went inside a store for shelter. He listened to the radio warning about a possible tornado coming. He stood by the window to watch the lightning. Then, he went with the store owner into the basement to wait for the tornado to pass.

SEP **Constructing Explanations and Designing Solutions**

Activity 14

Analyze Like a Scientist

Quick Code:
us3809s

Protecting Yourself from Severe Weather

Read the text and **watch** the video to learn more about staying safe during severe weather events. As you read, **underline** times when it is best to stay inside, and **circle** times when it is best to evacuate.

Protecting Yourself from Severe Weather

Severe storms are dangerous, but your actions can keep you safe. As much as possible, prepare for severe weather in advance. Make an emergency kit with extra water, food, and first aid supplies. During the storm, listen carefully to warnings and instructions—these may come from adults you know, or from announcements on TV, radio, or the Internet. Avoid dangerous activities such as going outside to see lightning or a **tornado**. When necessary, take action. The action you need to take will depend on the type of severe weather.

Emergency Shelter

Wind: During high winds caused by tornadoes or hurricanes, you should move to the basement or to a room without windows. This protects you from flying debris and glass from broken windows. Never stay in a mobile home during a tornado because some tornadoes are strong enough to knock over mobile homes.

Flood waters: During a hurricane or other extreme precipitation event, you may need to move to higher ground to avoid flooding. Avoid fast-moving waters and do not cross flooded roads.

Lightning: Lightning is a dangerous form of electricity. It often strikes the highest object in an area, and it travels easily through metal. During a thunderstorm, you should move away from open fields and trees. Take shelter in a covered area such as a house or a school.

Storm damage: Hurricanes like Hurricane Harvey in 2017 can damage buildings, water supplies, and power plants. After the storm, it may be nearly impossible to find clean water, electricity, food, or clean clothes. By having a family emergency plan and a kit for severe weather, you will be prepared to survive for a few days without these items.

Protecting Yourself from Severe Weather *cont'd*

Landslide: The best response is to evacuate the area if there is time. This is because landslides can damage buildings and plants. During a landslide, listen for unusual sounds that might come from moving debris. Stay away from bridges, streams, rivers, and low areas. If you are caught in a landslide, curl into a tight ball and protect your head.

Wildfires: It may be necessary to evacuate your home if a wildfire is approaching. If there is time, you can remove any dry plants and fuel sources (such as a stack of wood or a propane grill) from around your house. Keep plants watered to make them less likely to catch on fire.

Severe Storm Safety

Activity 15

Evaluate Like a Scientist

Quick Code:
us3810s

Take Shelter

Using the space provided, **describe** the best place to find shelter during a hurricane or tornado. **Explain** your reasoning.

SEP **Obtaining, Evaluating, and Communicating Information**

Activity 16

Investigate Like a Scientist

Quick Code:
us3811s

Hands-On Investigation: Emergency Kit and Plan

In this activity, you will design a severe weather emergency kit and plan. After you design your kit and plan, you will bring it home to build an emergency kit and discuss the plan with your family.

Make a Prediction

Question	Prediction
What items should be included in a severe weather emergency kit?	
What is important in a severe weather emergency plan?	

What materials do you need? (per group)

- Magazine
- Scissors
- Paper
- Glue

What Will You Do?

1. Brainstorm items that should be included in a severe weather emergency kit. Fill in the graphic organizer to explain what each item is used for, why it is important, and where it can be found. Rank each item in order of importance.
2. Find images of the items you want in your emergency kit in magazines. Cut out the images and paste them into an emergency kit collage.
3. Discuss ideas for emergency plans with your group, including where to go, an escape route from your home, and who to call.
4. Write a draft of your emergency plan.
5. Take your collage and emergency plan draft home. Assemble the kit with your family and take a picture. Discuss your emergency plan and come up with a final draft as a family. Bring the picture and your final draft back to school to share with the class.

Item	What is the item used for?	Why is the item important during an emergency?	Where can you find this item?	Rank 1: Essential 2: Important 3: Nice to have

Think About the Activity

What did your emergency kit and plan look like before discussing it with your family? What did they look like after your family discussion? **Write** a simplified version of each in the Before and After sections of the graphic organizer. **Explain** why you made the changes you made in the Changes section.

Topic: **Emergency Kit and Plan**

Before:	After:

Changes:

Activity 17

Record Evidence Like a Scientist

Quick Code: us3812s

Mudslide

Now that you have learned about severe weather, look again at the Mudslide image. You first saw this in Wonder.

Let's Investigate Mudslide

Talk Together

How can you describe mudslides now? How is your explanation different from before?

SEP **Constructing Explanations and Designing Solutions**

Look at the Can You Explain? question. You first read this question at the beginning of the lesson.

Can You Explain?

How do safety actions protect people in different kinds of severe weather?

Now, you will use your new ideas about severe weather to answer a question.

1. **Choose** a question. You can use the Can You Explain? question or one of your own. You can also use one of the questions that you wrote at the beginning of the lesson.

2. Then, use the graphic organizer on the next page to help you answer the question.

 How do safety actions protect people in different kinds of severe weather?

My Question

My Claim

Evidence I Found

My Claim Is True Because

STEM in Action

Quick Code: us3813s

Activity 18

Analyze Like a Scientist

Skyscrapers Push the Limits in Storms

Read the text about skyscrapers, and then **complete** the activities that follow.

Skyscrapers Push the Limits in Storms

You have learned the importance of staying safe during storms. Structural engineers design buildings and other structures that can withstand the power of severe weather. They design building models, test them, and redesign them to make sure the buildings are safe, even in severe weather.

In Tokyo, Japan, there are a lot of tall buildings. Strong winds can be very dangerous to these buildings. Watch the video segment to learn how engineers work to help keep these buildings safe.

Obtaining, Evaluating, and Communicating Information

Like the engineers in Japan, structural engineers in the United States have important jobs. They also use engineering processes to make sure that structures are safe. This is especially important when severe weather happens. Structures must be able to hold up against dangerous forces like heavy winds and powerful floods.

Hurricane-Safe Home

Imagine you are a structural engineer. You have been asked to design a home that will remain as safe as possible during hurricanes. **Describe** two characteristics of the home that you would like to include in your design to keep the home as safe as possible during hurricanes. Use scientific reasoning to support your answer.

Design for Coastal Living

The home in this image is located along the coast. **Identify** a structure on the home that was likely designed to help keep the home safe during floods. **Explain** your reasoning.

Boathouse

Activity 19

Evaluate Like a Scientist

Quick Code: us3814s

Review: Weather Hazards

Think about what you have read and seen. What did you learn?

Write down some key ideas you have learned. **Review** your notes with a partner. Your teacher may also have you take a practice test.

Talk Together

Think about what you saw in Get Started. Use your new ideas to discuss how you can prepare to be safe during severe weather.

SEP **Obtaining, Evaluating, and Communicating Information**

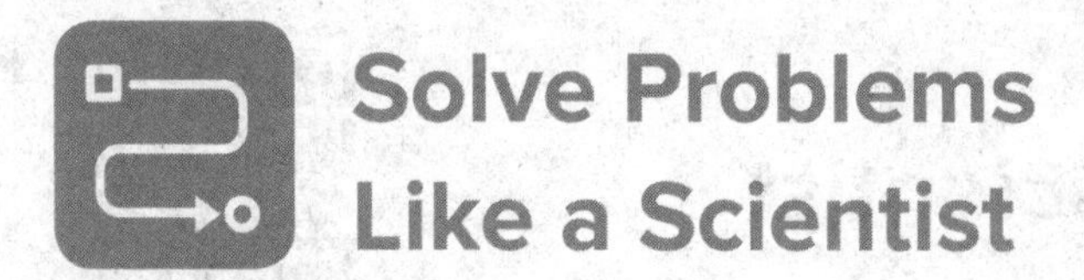

Solve Problems Like a Scientist

Quick Code:
us3816s

Unit Project: Mudslides and Floods

In this project, you will use what you know about weather to design solutions to flooding and mudslides.

Floods often come when an area gets too much rain. If the soil is too dry or rocky, or is already very wet, the rain will run off quickly into a lake, river, stream, or ocean. As this happens, the water rises in the local lake, river, stream, or ocean. Homes and lives can be in danger if the water rises too rapidly. When floods approach, people often make walls of sandbags to keep the water back. Even though the wall of sandbags may let some water through, it holds enough water back to keep the areas behind it dry.

Mudslides also can occur when an area gets too much rain. If the soil on hillsides is eroded or bare, the rain can turn the soil to mud. Then, the mud may flow down the hill. Mudslides are most dangerous to homes and people at the base of a hill, but they can flow onto roads and beyond.

To address these issues, in this project, you will design a barrier to prevent either floods or mudslides.

Mudslides and Floods

SEP Engaging in Argument from Evidence
SEP Asking Questions and Defining Problems
SEP Constructing Explanations and Designing Solutions
CCC Cause and Effect

Design Barrier

Recall the impacts of floods and mudslides. **Think** about the requirements for a barrier that can prevent either floods or mudslides.

The goal of your design is to stop water or mud from flowing to a certain area.

What do you need to consider when designing your barrier?

How will your design work to prevent damage from flooding or mudslides?

What materials will you need for your design?

How can you test your design?

Diagram

Sketch a diagram of your barrier. **Label** all the parts of your design.

Think About the Activity

Build a model of your barrier. Then, **test** your design. **Write** or **draw** your answers to the questions in the chart.

What Worked?	What Did Not Work?

What Could Work Better?

Grade 3 Resources

- Bubble Map
- Safety in the Science Classroom
- Vocabulary Flash Cards
- Glossary
- Index

Name ____________________

Bubble Map

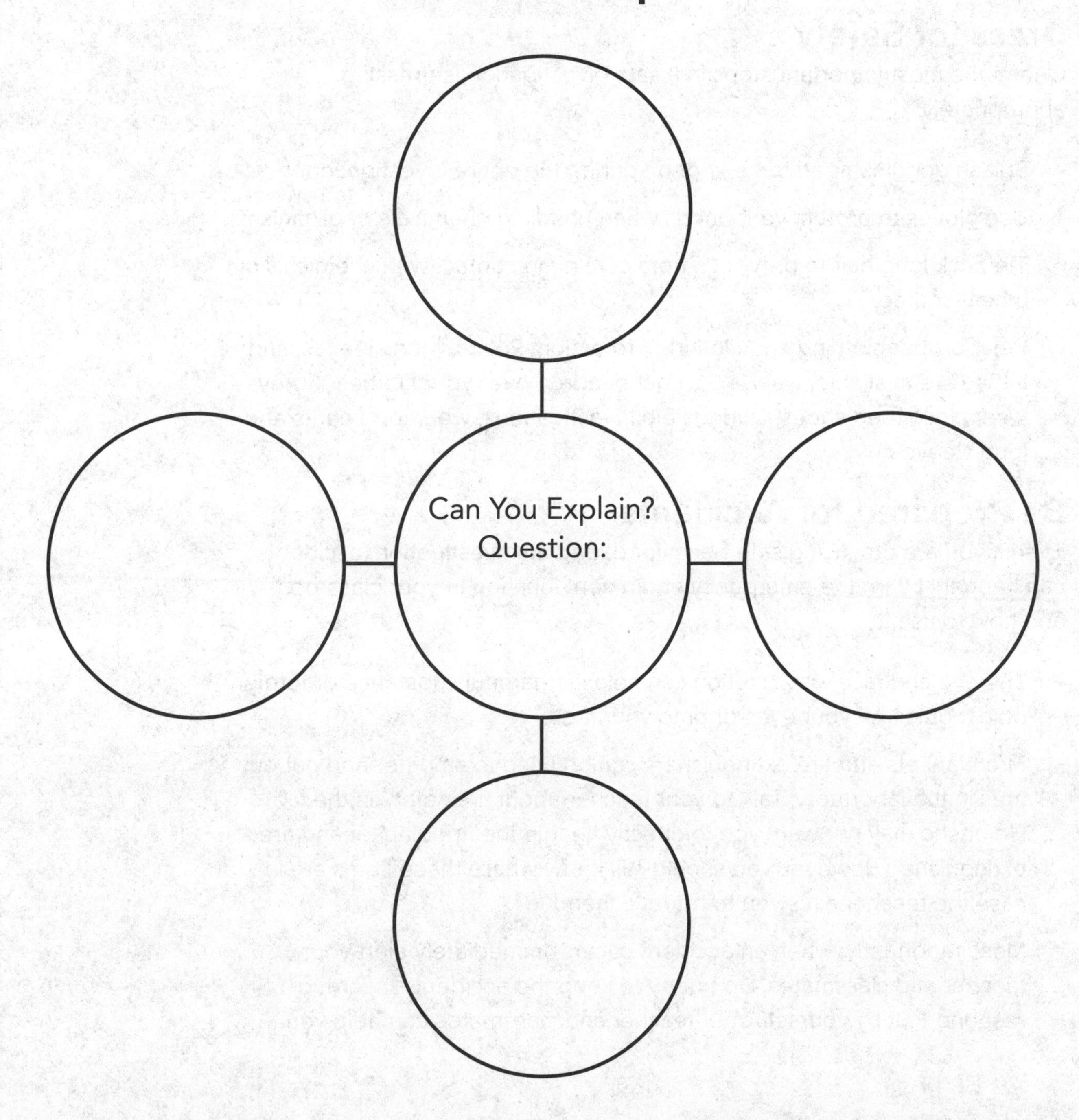

Safety in the Science Classroom

Following common safety practices is the first rule of any laboratory or field scientific investigation.

Dress for Safety

One of the most important steps in a safe investigation is dressing appropriately.

- Splash goggles need to be kept on during the entire investigation.
- Use gloves to protect your hands when handling chemicals or organisms.
- Tie back long hair to prevent it from coming in contact with chemicals or a heat source.
- Wear proper clothing and clothing protection. Roll up long sleeves, and if they are available, wear a lab coat or apron over your clothes. Always wear close toed shoes. During field investigations, wear long pants and long sleeves.

Be Prepared for Accidents

Even if you are practicing safe behavior during an investigation, accidents can happen. Learn the emergency equipment location in your classroom and how to use it.

- The eye and face wash station can help if a harmful substance or foreign object gets into your eyes or onto your face.
- Fire blankets and fire extinguishers can be used to smother and put out fires in the laboratory. Talk to your teacher about fire safety in the lab. He or she may not want you to directly handle the fire blanket and fire extinguisher. However, you should still know where these items are in case the teacher asks you to retrieve them.
- Most importantly, when an accident occurs, immediately alert your teacher and classmates. Do not try to keep the accident a secret or respond to it by yourself. Your teacher and classmates can help you.

Practice Safe Behavior

There are many ways to stay safe during a scientific investigation. You should always use safe and appropriate behavior before, during, and after your investigation.

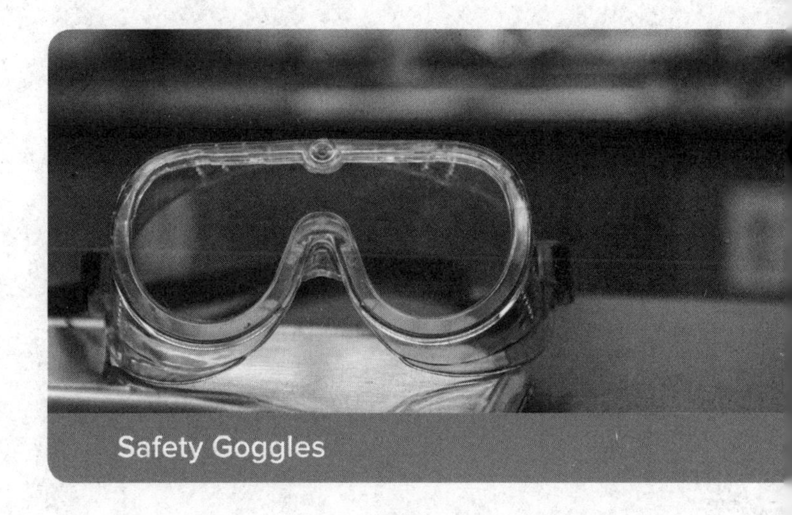
Safety Goggles

- Read the all of the steps of the procedure before beginning your investigation. Make sure you understand all the steps. Ask your teacher for help if you do not understand any part of the procedure.
- Gather all your materials and keep your workstation neat and organized. Label any chemicals you are using.
- During the investigation, be sure to follow the steps of the procedure exactly. Use only directions and materials that have been approved by your teacher.
- Eating and drinking are not allowed during an investigation. If asked to observe the odor of a substance, do so using the correct procedure known as wafting, in which you cup your hand over the container holding the substance and gently wave enough air toward your face to make sense of the smell.
- When performing investigations, stay focused on the steps of the procedure and your behavior during the investigation. During investigations, there are many materials and equipment that can cause injuries.
- Treat animals and plants with respect during an investigation.
- After the investigation is over, appropriately dispose of any chemicals or other materials that you have used. Ask your teacher if you are unsure of how to dispose of anything.
- Make sure that you have returned any extra materials and pieces of equipment to the correct storage space.
- Leave your workstation clean and neat. Wash your hands thoroughly.

air pressure

Image: Hans Braxmeier/Pixabay

the force that air puts on an area

atmosphere

Image: Paul Fuqua

layers of gas that surround a planet

barometer

Image: Mtsaride / Shutterstock.com

a tool used to measure air pressure

climate

Image: Discovery Communications, Inc.

the average weather conditions in an area

coast

Image: Paul Fuqua

an area where the ocean meets the land

data

Image: Paul Fuqua

measurements or observations

detect

Image: Discovery Communications, Inc.

to notice or find, often with the help of a science tool

drought

Image: R_Tee / Shutterstock.com

a prolonged shortage of rainfall

equator

Image: NASA

an imaginary line located halfway between the North and South Poles

forecast

Image: RitaE/Pixabay

(n) a prediction about what the weather will be like in the future

heat

Image: Paul Fuqua

what happens when thermal energy is gained or lost

humidity

Image: Paul Fuqua

the measure of how much water vapor is in the air

hurricane

Image: Discovery Education

a storm with strong winds and rain that forms over tropical waters

lightning

Image: NOAA

happens when electric currents flow between a cloud and the ground or between two clouds

meteorology

Image: Brin Weins/Pixabay

the study of patterns of weather

precipitation

Image: Paul Fuqua

water that is released from clouds in the sky; includes rain, snow, sleet, hail, and freezing rain

predict

Image: FrameStockFootages / Shutterstock.com

to make a guess based on what you already know

rain

Image: Paul Fuqua

liquid water that falls from the sky

region

Image: Discovery Education

places, especially around the world

severe

Image: Johnny Lindner/Pixabay

dangerous or harsh conditions

tornado

Image: John D Sirlin / Shutterstock.com

a funnel-shaped cloud that rotates at high speeds and extends down from a cloud to the ground

water

Image: Paul Fuqua

a compound made of hydrogen and oxygen; can be in either a liquid, ice, or vapor form and has no taste or smell

weather

Image: Discovery Communications, Inc.

the properties of the atmosphere at a given time and location, including temperature, air movement and precipitation

wind

Image: Discovery Communications, Inc.

the movement of air due to atmospheric pressure differences

English — A — Español

adapt
something a plant or animal does to help it survive in its environment

adaptarse
algo que una planta o animal hace para sobrevivir en su medio ambiente

adaptation
how a plant or animal has changed over time to help it survive in its environment (related word: adapt)

adaptación
manera en la que ha cambiado una planta o un animal con el transcurso del tiempo para sobrevivir en su medio ambiente (palabra relacionada: adaptar)

adjust
to change one's position or behavior to allow for a better fit, to adapt

acomodarse
cambiar de posición o comportamiento para ajustarse mejor, adaptarse

air
a gas that is all around you and you can't see it, but living things like plants and animals need it to breathe and to grow

aire
gas que nos rodea y que no podemos ver, pero que las plantas y los animales necesitan para respirar y crecer

air pressure
the force that air puts on an area (related word: pressure)

presión de aire
fuerza que el aire ejerce sobre un área (palabra relacionada: presión)

---

analyze
to closely examine something and then explain it

analizar
examinar con atención algo y luego explicarlo

---

ancient
very old

antiguo
extremadamente viejo

---

arctic
being from an icy climate, such as the North Pole

ártico
que pertenece a un clima helado, como el Polo Norte

---

artificial selection
specifically breeding animals or cultivating plants only for certain desired genetic outcomes

selección artificial
criar animales o cultivar plantas específicamente para obtener determinados resultados genéticos deseados

atmosphere
layers of gas that surround a planet (related word: atmospheric)

atmósfera
capas de gas que rodean un planeta (palabra relacionada: atmosférico)

attract
to pull one thing toward another (related word: attraction)

atraer
jalar un objeto hacia otro (palabra relacionada: atracción)

B

balanced forces
when two equal forces are applied to an object in opposite directions, the object does not move

fuerza equilibrada
cuando se aplican dos fuerzas iguales sobre un objeto en direcciones opuestas, el objeto no se mueve

barometer
a tool used to measure air pressure (related word: barometric)

barómetro
herramienta usada para medir la presión del aire (palabra relacionada: barométrico)

behavior
the way in which a living thing acts (related word: behave)

conducta
manera en la que actúa un ser vivo (palabra relacionada: comportarse)

C

camouflage
the coloring or patterns on an animal's body that allow it to blend in with its environment

camuflaje
color o patrones del cuerpo de un animal que le permite mezclarse con su medioambiente

---

carnivore
a meat eater

carnívoro
que se alimenta de carne

---

characteristic
a special quality that something may have

característica
cualidad especial de algo

---

climate
the usual weather conditions in a place or area (related word: climatic)

clima
condiciones del tiempo atmosférico habituales en un lugar o área (palabra relacionada: climático)

coast
an area where the ocean meets the land

costa
área donde el océano se encuentra con la tierra

community
a group of different populations that live together and interact in an environment

comunidad
grupo de distintas poblaciones que viven juntas e interactúan en un ambiente

contact
when two things are so close they touch

contacto
cuando dos objetos están tan cerca que se tocan

coral reef
an area that forms in the warm, shallow ocean waters made from the hard skeletons of animals called corals

arrecife de coral
área que se forma en aguas marinas cálidas y poco profundas a partir del esqueleto duro de animales llamados corales

cycle
a process that repeats (related word: cyclical)

ciclo
proceso que se repite (palabra relacionada: cíclico)

D

data
measurements or observations (related word: datum)

datos
medidas u observaciones (palabra relacionada: dato)

---

desert
an area that gets very little rain water and does not have a lot of growing plants

desierto
área que recibe muy poca precipitación y tiene muy poca vegetación

---

detect
to notice or find, often with the help of a science tool (related words: detection, detector)

detectar
notar o encontrar, generalmente con la ayuda de una herramienta científica (palabras relacionadas: detección, detector)

---

dinosaur
an extinct organism with reptile and birdlike features: Dinosaurs lived on Earth millions of years ago

dinosaurio
organismo extinto con características de reptil y ave: los dinosaurios vivían en la Tierra hace millones de años

---

discharge
the release of energy

descarga
liberación de energía

drought
a long period of little or no rain

sequía
escasez o ausencia prolongada de lluvia

ecosystem
all the living and nonliving things in an area that interact with each other

ecosistema
todos los seres vivos y objetos sin vida de un área, que se interrelacionan entre sí

electrical charges
a type of charge, either positive, negative, or neutral

carga eléctrica
un tipo de carga, ya sea positiva, negativa, o neutra

electrical energy
energy produced by power plants that flows through electrical lines and wires

energía eléctrica
energía producida por centrales eléctricas que fluye a través de cables y líneas eléctricas

electromagnet
a metal object that acts as a magnet when an electric current moves through it

electroimán
objeto de metal que actúa como un imán cuando una corriente eléctrica pasa a través de él

endangered a type of plant or animal that is in danger of becoming extinct	**amenazado** tipo de planta o animal que está en peligro de extinción
energy the ability to do work or make something change	**energía** capacidad para hacer un trabajo o producir un cambio
environment all the living and nonliving things that surround an organism	**medio ambiente** todos los seres vivos y objetos sin vida que rodean a un organismo
equator an imaginary line that divides Earth into Northern and Southern Hemispheres; located halfway between the North and South Poles (related word: equatorial)	**ecuador** línea imaginaria que divide la Tierra en Hemisferio Norte y Hemisferio Sur; ubicada a mitad de camino entre el Polo Norte y el Polo Sur (palabra relacionada: ecuatorial)

evidence
facts that give us more information, clues, or proof about something else

evidencia
hechos que nos dan más información, pistas o pruebas sobre otra cosa

---

extinct
when a plant or an animal is no longer in existance (related word: extinction)

extinto
cuando una planta o un animal ya no existe (palabra relacionada: extinción)

F

factor
something that influences another thing to move or change

factor
algo que influye en que otra cosa se mueva o cambie

---

food web
a model that shows many different feeding relationships among living things

red alimentaria
modelo que muestra muchas y diferentes relaciones de alimentación entre los seres vivos

force
a pull or push that is applied to an object

fuerza
acción de atraer o empujar que se aplica a un objeto

---

forecast
(v) to analyze weather data and make an educated guess about weather in the future; (n) a prediction about what the weather will be like in the future based on weather data

pronosticar / pronóstico
(v) analizar los datos del tiempo y hacer una conjetura informada sobre el tiempo en el futuro; (s) predicción sobre cómo será el tiempo en el futuro en base a datos

---

fossil
the remains of a living animal or plant from a very long time ago (related word: fossilize)

fósil
restos de un animal o planta de hace mucho tiempo (palabra relacionada: fosilizar)

---

friction
when two objects rub against each other

fricción
cuando dos objetos se frotan entre sí

generation
the next group of living things or species that will be born around the same time

generación
el siguiente grupo de seres vivos o especies que nacerán alrededor de la misma época

---

germination
when a young plant grows from a seed (related word: germination)

germinación
proceso por el cual una planta joven brota de una semilla (palabra relacionada: germinar)

---

grassland
a large area of land covered by grass

pradera
gran área de tierra cubierta principalmente de hierba

---

gravity
the force that pulls an object toward the center of Earth (related word: gravitational)

gravedad
fuerza que jala a un objeto hacia el centro de la Tierra (palabra relacionada: gravitacional)

H

habitat
the place where a plant or animal lives

hábitat
lugar donde vive una planta o un animal

heat
a form of energy; the state of being very warm

calor
transferencia de energía térmica

herbivore
a plant eater

herbívoro
que se alimenta de vegetales

humidity
the measure of how much water vapor is in the air

humedad
medida de cuánto vapor de agua hay en el aire

hurricane
a storm with strong winds and rain that forms over tropical waters (related terms: typhoon, tropical cyclone)

huracán
tormenta con fuertes vientos y lluvia que se forma sobre aguas tropicales (palabras relacionadas: tifón, ciclón tropical)

I

impact
to influence or affect something

impactar
afectar o influir en algo

inherit
to get genetic information and traits from a parent or parents (related word: inheritance)

heredar
obtener información y rasgos genéticos de uno o ambos padres (palabra relacionada: herencia)

instinct
behaviors animals and people are born with that help them survive

instinto
conductas con las que nacen los animales y las personas y que los ayudan a sobrevivir

interact
to act on one another (related word: interaction)

interactuar
ejercer influencia mutua (palabra relacionada: interacción)

L

life cycle
the various stages of an organism's development and reproduction

ciclo de la vida
diversas etapas del desarrollo y de la reproducción de un organismo

lifespan
how long in time an organism is expected to live

longevidad
cuánto tiempo se espera que viva un organismo

---

lightning
when electricity flows between a cloud and the ground or between two clouds and you sometimes see a streak or a flash in the sky

relámpago
cuando fluye electricidad entre una nube y el suelo o entre dos nubes, y a veces se ve una raya o un destello en el cielo

M

magnetic
having the properties of a magnet; having the ability to be attracted to or by a magnet

magnético
que tiene las propiedades de un imán; que tiene la capacidad de ser atraído hacia o por un imán

---

magnetic field
a region in space near a magnet or electric current in which magnetic forces can be detected

campo magnético
región en el espacio cerca de un imán o de una corriente eléctrica, donde pueden detectarse fuerzas magnéticas

magnetism
the amount of attraction to a magnet

magnetismo
la cantidad de atracción hacia un imán

---

mature
when a living thing is fully grown or an adult (related word: maturity)

maduro
organismo que ha crecido por completo, o que es adulto (palabra relacionada: madurez)

---

metamorphosis
when a living thing goes through changes during its life cycle, like a frog or a butterfly

metamorfosis
cuando un ser vivo experimenta cambios durante su ciclo de vida, como en el caso de las ranas o las mariposas

---

meteorology
the study of patterns of weather

meteorología
estudio de los patrones del tiempo atmosférico

---

microorganisms
the tiniest of living things, can only be seen under a microscope

microorganismo
los seres vivos más diminutos, que sólo se pueden ver con un microscopio

migrating
traveling within a group to a different location during season changes

migratorio
que viaja dentro de un grupo a un lugar diferente durante los cambios de estación

motion
when something moves from one place to another (related terms: move, movement)

movimiento
cuando algo pasa de un lugar a otro (palabras relacionadas: mover, desplazamiento)

N

natural
not human-made (related word: nature)

natural
que no está hecho por un ser humano (palabra relacionada: naturaleza)

negative charge
a charge that you get when there is a build-up of electrons

carga negativa
carga que resulta de la acumulación de electrones

neutral
having no electrical charge, being neither positive nor negative

neutro
que no tiene carga eléctrica, que no es positivo ni negativo

nutrient
something in food that helps people, animals and plants live and grow

nutriente
algo en los alimentos que ayuda a las personas, los animales y las plantas a crecer

O

observe
to study something using your senses (related word: observation)

observar
estudiar algo usando tus sentidos (palabra relacionada: observación)

offspring
a new organism that is produced by one or more parents

descendencia
organismo nuevo originado por uno o más progenitores

organism
any individual living thing

organismo
todo ser vivo individual

P

parasite a plant, animal, or fungus that lives on or in another living thing to get food and energy from it	**parásito** una planta, animal, u hongo que vive sobre o dentro de otro ser vivo, del cual se alimenta y obtiene energía
pendulum a string or bar that is loose at one end but fixed at the other end and can swing back and forth, like in a clock	**péndulo** cuerda o barra que posee un extremo suelto y otro fijo y puede balancearse de un lado a otro como en un reloj
pole the opposite ends of a battery, a magnet, or the north and south ends of Earth	**polo** extremos opuestos de una batería, un imán, o los extremos norte y sur de Tierra
pollen the yellow powder found inside of a flower (related word: pollinate)	**polen** polvo amarillo que se encuentra dentro de una flor (palabra relacionada: polinizar)

pollution
when harmful materials have been put into the air, water, or soil (related word: pollute)

contaminación
cuando se introducen materiales perjudiciales en el aire, el agua o el suelo (palabra relacionada: contaminar)

---

positive charge
a charge that you get when there are more protons than electrons

carga positiva
carga que resulta cuando hay más protones que electrones

---

precipitation
water that is released from clouds in the sky; includes rain, snow, sleet, hail, and freezing rain

precipitación
agua liberada de las nubes en el cielo; incluye la lluvia, la nieve, la aguanieve, el granizo, y la lluvia congelada

---

predators
the larger animals that hunt the smaller animals, or prey, for food

depredador
animales más grandes que cazan a otros más pequeños, o presas, para alimentarse

---

predict
to make a guess based on what you already know (related word: prediction)

predecir
hacer una suposición a partir de lo que ya se sabe (palabra relacionada: predicción)

prehistoric a time before history was written	**prehistórico** época antes de que se escribiera la historia
prey the animals that get hunted by the larger animals, or predators, for food	**presa** animales que son cazados por animales más grandes, o depredadores, como alimento

R

rain liquid water that falls from the sky	**lluvia** agua líquida que cae desde el cielo
recycle to create new materials from used products	**reciclar** crear nuevos materiales a partir de productos usados
region a place, especially around the world	**región** lugar, especialmente alrededor del mundo

repel
to force an object away or to keep it away

repeler
forzar a un objeto para que se aleje o mantenerlo alejado

reproduce
to make more of a species; to have offspring (related word: reproduction)

reproducir
engendrar más individuos de una especie; tener descendencia (palabra relacionada: reproducción)

S

seed
the small part of a flowering plant that grows into a new plant

semilla
parte pequeña de una planta con flor que crece y se convierte en una nueva planta

seedling
a young plant that grows from a seed

plántula
planta joven que crece de una semilla

severe
dangerous or harsh conditions

severo
condiciones peligrosas o adversas

species
a group of the same kinds of living things

especie
un grupo de las mismas clases de seres vivos

---

static electricity
electric charges that build up on an object

electricidad estática
cargas eléctricas que se acumulan sobre un objeto

---

stored energy
energy in an object or substance that is not being given off by the object or substance

energía almacenada
energía en un objeto o una sustancia que no es liberada por el objeto o la sustancia

---

survival
ability to live and remain alive

supervivencia
capacidad de vivir y mantenerse vivo

---

survive
to continue living or existing: an organism survives until it dies; a species survives until it becomes extinct (related word: survival)

sobrevivir
continuar viviendo o existiendo: un organismo sobrevive hasta que muere; una especie sobrevive hasta que se extingue (palabra relacionada: supervivencia)

T

temperature (general) a measure of how hot or cold a substance is	**temperatura (general)** medida de cuán caliente o fría es una sustancia
tornado a funnel-shaped cloud or column of air that rotates at high speeds and extends downward from a cloud to the ground	**tornado** nube o columna de aire con forma de embudo que rota a altas velocidades y se extiende hacia abajo desde una nube hasta el suelo
trait a characteristic that you get from one of your parents	**rasgo** característica que se obtiene de uno de los progenitores
tropical from a warmer climate, near the equator	**tropical** que pertenece a un clima más cálido, cerca del ecuador

W

water a clear liquid that has no taste or smell	**agua** líquido transparente que no tiene sabor ni olor

weather the properties of the atmosphere at a given time and location, including temperature, air movement, and precipitation	**tiempo atmosférico** propiedades de la atmósfera en un determinado momento y lugar; entre ellas, la temperatura, el movimiento, de aire y las precipitaciones
wind the movement of air due to atmospheric pressure differences	**viento** movimiento de aire que se produce por las diferencias en la presión atmosférica
work a force applied to an object over a distance	**trabajo** fuerza aplicada a un objeto a lo largo de una distancia

Index

O

P

R

S

T

U

W